ENDURING LEGACIES

PEOPLE OF THE 1926 BARNES-HECKER MINE DISASTER

MARY V. TIPPETT

Foreword by Troy Henderson

Modern History Press

Ann Arbor, MI

Enduring Legacies: People of the 1926 Barnes-Hecker Mine Disaster
Copyright © 2025 by Mary V. Tippett. All Rights Reserved.

ISBN 979-8-89656-030-2 paperback
ISBN 979-8-89656-031-9 hardcover
ISBN 979-8-89656-032-6 eBook

Modern History Press   www.ModernHistoryPress.com
5145 Pontiac Trail   info@ModernHistoryPress.com
Ann Arbor, MI 48105
Tollfree 888-761-6268   FAX 734-663-5851

Distributed by Ingram Group (USA, CAN, AU, UK, EU)

Library of Congress Cataloging-in-Publication Data

Names: Tippett, Mary V., 1950- author
Title: Enduring legacies : the people of the 1926 Barnes-Hecker Mine
    Disaster / Mary V. Tippett.
Other titles: People of the 1926 Barnes-Hecker Mine Disaster
Description: Ann Arbor, MI : Modern History Press, [2025] | Includes
    bibliographical references and index. | Summary: "An unprecedented
    presentation of the people and milieu of the 51 men lost in
Michigan's
    worst mining disaster in November 1926. Based on oral histories of
    descendants and family members, rare documents and photos,
    archival records, and contemporaneous news reports"-- Provided by
publisher.
Identifiers: LCCN 2025024146 (print) | LCCN 2025024147 (ebook) |
ISBN 9798896560302 paperback | ISBN 9798896560319 hardcover |
ISBN 9798896560326 epub
Subjects: LCSH: Barnes-Hecker Mine Disaster, Mich., 1926 | Mine
    accidents--Michigan--Upper Peninsula | Cleveland-Cliffs Iron
Company | Iron mines and mining--Michigan--Upper Peninsula--
History--20th century | Upper Peninsula (Mich.)--Biography | Upper
Peninsula   (Mich.)--Genealogy | Upper Peninsula (Mich.)--History
Classification: LCC TN403.M5 T57 2025 (print) | LCC TN403.M5
(ebook)
LC record available at https://lccn.loc.gov/2025024146
LC ebook record available at https://lccn.loc.gov/2025024147

"We are not forgetting the families of these men."

K.I. Sawyer, Mayor of Ishpeming, Michigan

Chicago Daily Tribune, November 5, 1926

I shall pass through this world but once.
Any good, therefore, that I can do
Or any kindness I can show
to another human being,
Let me do it now.
Let me not postpone it or neglect it.
For I shall not pass this way again.

—Anonymous

Contents

Acknowledgements

In addition to the organizations and people already mentioned throughout the book, special thanks are also due to:

- Troy Henderson and Barry James, historians at the Michigan Iron Industry Museum. Without their support and encouragement, this book would never have been written.

- Georgann Coon, retired language arts teacher, who volunteered her editing expertise.

- William (Bill) P. Kosonen, nature photographer and photo restoration.

- Linda Talsma, Marquette County clerk, and Tanya Nelson, for verifying many vital records.

Foreword

In the spring of 1999, I was a newly minted graduate of Northern Michigan University with a major in history. I was hired that summer to work a seasonal job at the Michigan Iron Industry Museum in Negaunee before moving away to graduate school. On my very first day on the job, I was asked by a visitor where the Barnes-Hecker monument was located on the museum grounds. I was relieved; I didn't know a lot about the job yet to field general questions, but I did know that. Growing up on the Marquette Iron Range, it was hard not to know about the Barnes-Hecker mine disaster of 1926. It was embedded in our school curriculum and community. My supervisor at the museum, Tom Friggens, had written a book on the tragedy called *No Tears in Heaven: The 1926 Barnes-Hecker Mine Disaster*.

After completing graduate school in Chicago, I returned to Negaunee and took a job as a historian with the Michigan History Center, headquartered at the same museum I worked at a dozen years earlier.

In the spring of 2016, received a call from Mary V. Tippett. I knew Mary was a descendant of a miner who died in the Barnes-Hecker, and I knew of her dedicated research and oral histories of local miners. Mary wanted to bring her dedication to form a Remembrance Committee to commemorate the 90th anniversary. I joined her team, but I certainly could not match her energy and devotion to the remembrance.

The culmination of the 90th remembrance in 2016 was a series of events and services that focused on legacy. The ripple effect of such a heartbreaking event in a small community now spanned generations, and people from around the country and indeed globe whose lives were affected by the tragedy. In the wake of the emotional remembrance events, Mary recounted numerous stories and connections she had made to me and my colleague Barry James. After listening and absorbing the poignant stories, Barry remarked "You've got a book there."

Buoyed by the achievements of 2016, Mary continued her outreach to gather stories from descendants. Untold hours of her oral histories, conversations, emails, and research formed the framework of this book. Photographs of mining families and their descendants that Mary collected breathe life into the narrative. After Mary shared some early draft chapters with me, I asked her who she felt the audience for this book was. She responded that it was for the descendants of the Barnes-Hecker mine disaster and for the many people whose lives were impacted by it.

Enduring Legacies: People of the 1926 Barnes-Hecker Mine Disaster presents the stories from descendants in a rich collection that reveals what happened to the widows, parents, siblings, and children after the sudden death of 51 men in a small and close-knit community. How did they cope in the days after the cave-in? How did they move on in the months and years after a chair at their dinner table remained vacant? How would they relay the stories of their loved ones to their grandchildren and great-grandchildren? How did the memories of the 51 men, most of which are still entombed in the abandoned iron mine, weave through generations of people?

This book is about a shared experience, and it reveals that it was shared by many. Some of the chapters uncover the experience of people who did not lose a husband or father, but a friend, a fellow miner, or a member of a community. This book displays a support structure for a devastated community. Taken as a whole, it shows readers that the bonds between people affected by the disaster, whether it be in 1926 or 2026, make the legacy of the Barnes-Hecker an expanding story.

Troy Henderson
Historian, Michigan History Center
Michigan Iron Industry Museum

Introduction

Of cardboard boxes, echoes, and grasshoppers

When 51 men perished in the Barnes-Hecker Mine disaster of November 3, 1926, they left dark clouds of mourning and sleepless nights for the people left behind. And for many people connected with the tragedy, wisps of those dark clouds have lingered through almost 100 years.

What happened to the people left behind? That is the premise for this book: finding out what happened to the widows and families… and the families of the men who were not lost.

There was a pent-up longing for the family stories to be told. What you'll read is what the descendants have had to say.

As the 100th anniversary of the disaster approaches, the legacies of tragedy are revealed in the stories of the people… a widow whose luggage consisted of cardboard boxes; a youth who climbed into the abandoned shaft to make echoes; and fond memories of a stern but humorous grandfather who said he could only grow boys and grass-hoppers.

The experiences and expressions of loss are as varied as the families themselves, and yet their commonalities bind them together in a kinship of loss, triumph, survival, and love.

Preface & Dedication

Dateline: Wednesday, November 3, 1926, 11:30 AM

Catastrophe descended on the Barnes-Hecker Mine near Ishpeming, Michigan, when 200,000 cubic yards of earth, water and debris collapsed into the underground workings of the Barnes-Hecker.

Fifty-one men perished, leaving 41 widows, 132 minor children, an unknown number of children and grandchildren, along with grieving parents, siblings, loved ones and co-workers. The tragedy left life-long scars on the other 90 men who were either safe on the surface or not on duty when the disaster happened.

The tragedy remains the worst mine disaster in Michigan history, and the worst iron ore mine disaster in the history of the United States.

The town of Ishpeming is located about 20 miles from the port city of Marquette on Lake Superior in Michigan's Upper Peninsula.

Our mission is to tell the stories of as many of the Barnes-Hecker families as possible. In our quest for names contained in original records and newspaper accounts, we have also extracted first-person stories of others involved in the incident.

The genesis for this book was the 90th anniversary Barnes-Hecker Remembrance in 2016 when hundreds of people attended events, and thousands followed the information posted on the Barnes-Hecker Remembrance Facebook page. We wanted to follow up with as many families as possible, and the response has been remarkable.

The research has included interviews with descendants of the Barnes-Hecker men and women; study of the original Cleveland-Cliffs Inc. (CCI) records and maps from the Central Upper Peninsula and Northern Michigan University Archives; and documents, photos, and videos in collections of the Archives of Michigan, Michigan Iron Industry Museum, Cliffs Shaft Mine Museum, Marquette Regional

History Center, private collections, and collections of the Brown County Library, Newberry Library, and the Library of Michigan. Records from Ancestry.com have also aided in clarifying some of the family relationships.

The entire digitized collection of audio interviews, photographs and documents collected through this project will be gifted to the Northern Michigan University Archives, Michigan Iron Industry Museum, Cliffs Shaft Mine Museum, Marquette Regional History Center, and Archives of Michigan so everything will be available for use by other researchers and future generations.

This book is dedicated to all the people whose lives have been swept up into the Barnes-Hecker Mine tragedy of November 3, 1926.

1

Catastrophe

The women were up early, stoking fires in kitchen wood stoves, making breakfast, packing lunch pails, and getting kids ready for school. Many of the men and older kids had morning chores before leaving for work and school.

Mary Valenti hadn't slept much, with four-month-old Nicholas waking every few hours for night-time feedings. Mary's husband Nick had to catch the "bus"—a canvas-covered truck—that came from the Barnes-Hecker Mine to Ishpeming to pick up men for the day shift. Breakfast was a busy time for the household of seven children.

In the nearby neighborhood of North Lake, Amanda Ranta was uncomfortably pregnant. She and her husband Elias talked about their son Reino who was in the hospital. After Elias left for work at the Barnes-Hecker, three of the older kids were bundled up for a short walk to North Lake School. When the morning hubbub slowed down, Amanda turned her attention to the two youngest children and her morning chores.

In the Barnes-Hecker *Location* (neighborhood) near the mine, 33 youngsters headed off to school, and the women's routines looked like a typical day.

Everything changed shortly after 11:30 that morning. It was November 3, 1926, and the morning routines would never be the same again. Shock and disbelief gave way to despair as word spread that none of the men trapped underground at the Barnes-Hecker could have survived a cave-in that left a crater the size of a city block over the workings of the mine.

The mine was operated by Cleveland-Cliffs Iron Company in Michigan's Upper Peninsula.

The Barnes-Hecker was about seven miles west of the city of Ishpeming, and twenty miles from the port city of Marquette on Lake Superior.

At 11:20 on the morning of November 3, a cave-in occurred, taking the lives of 51 men. Only one escaped the flooding mine alive. The catastrophe became the worst mine disaster in Michigan history, and the worst iron ore mining disaster in the history of the United States.

Forty-one women were widowed. One hundred and thirty-two minor children lost their dads. An unknown number of adult children, grandchildren, parents, siblings, loved ones, neighbors and co-workers were left to grieve the staggering loss of life. The remains of only ten men were recovered. Forty-one remain entombed in the mine.

[Author's note: the scenarios in the households are based on what is known about the families.]

On the evening of November 3, 1926, family members and co-workers of the lost men cluster at the Barnes-Hecker Mine. (Courtesy of Marquette Regional History Center and Michigan Iron Industry Museum)

2   The 51 Lost Men

In tribute to the lost men, their wives and families.

Barnes-Hecker Location:

- Peter DeRoche and his widow Evangeline
- Thomas Drew
- *Nels J. Hill and his widow Mae
- *Joseph Mankee
- Peter Mongiat and his widow Valentine
- Sam Phillippi (Sante Filippi) and his widow Mary
- John Santti and his widow Nanna (Juho Nestori Yli-Säntti)
- *William F. Tippett and his widow Nellie

Diorite:

- *John Arvi Wepsala

Ishpeming:

- Herman Aho and his widow Hilma
- Peter Carlyon and his widow Sarah Jane
- Raymond Carlyon (their son)
- William Carlyon and his widow Elly
- Edwin Chapman and his widow Elsie
- Herman Chapman (their son)
- Earl J. Ellersick and his widow Sarah
- Gust Franti and his widow Anna
- James Green and his widow Marie
- *Henry P. Haapala and his widow Ina

- *Jack Hanna and his widow Florence
- Arvid Heino and his bride Tyyne
- *William E. Hill and his widow Elizabeth
- Frank Jokinen
- John Arvid Kallio and his widow Sanna
- *Thomas Kirby Jr. and his widow Bertha
- *Thomas Kirby Sr. and his widow Anne
- Uno Koskinen and his widow Vieno
- Theodore Kiuru
- Ed Laituri and his widow Alle
- George Lampshire and his widow Alice
- Richard Lampshire and his late wife Lena
- John Luoma and his widow Ellen
- Emil Maki and his widow Lempi
- Walter Mattila (Ojaniemi) and his widow Lempi in Finland
- W. Harry Quayle and his widow Louisa
- Clement Simoneau and his widow Elna
- Erkki Timoharju (Ed Temo) and his widow Amalia in Finland
- Louis Trudell and his widow Delia
- William Tuomela his widow Florence
- Nick Valenti and his widow Mary
- Albert Wickman (Vickman)
- John Wiljanen and his widow Eliina

North Lake Location:

- Joseph Gelmi and his widow Emma
- *William Huot and his widow Clara
- William Kakkuri
- Elias Ranta and his widow Amanda
- James Scopel and his widow Frances
- Nestor Solomon (Sulonen) and his widow Matilda in Finland
- Solomon Walimaa and his widow Mary

West Ishpeming:

- Solomon Millimaki and his widow Minnie
- Walter Tippett and his widow Marian

*Men whose remains were recovered

[Author's note: The first list of names released to the press had 52 names, which included Frank Heino who was unable to report for work on the day of the tragedy. The *Daily Mining Journal* in Marquette published the list of 52 on the front page on November 4. There was also a case of mistaken identity when John Luoma was listed among the men recovered. Luoma's name was also listed in the November 6, 1926, *Iron Ore*; however, the correct identity was Nels Hill, and the papers published corrected lists. The tragic reality is that the bodies of several recovered men had sustained such severe injuries that they could only be identified by their clothing.].

3   Help for the Families

Stunned widows could not imagine how they would get through the night of the tragedy, let alone the vast unfocused void of the time that lay ahead. Neighbors and family members gathered in the women's homes with hot food, help with children, and comforting arms to wrap around the stricken widows.

Think about the nine women whose smallest children were 5 weeks old, 4 months, 8 months, 10 months, 11 months, and a year old. Five widows were pregnant. Florence Hanna gave birth to baby Jack on November 16. In the spring of 1927, Amanda Ranta gave birth to daughter Lillian; Vieno Koskinen had a son Stanley; Bertha Kirby had a son Tom; and Vangeline DeRoche had a daughter Ethel.

Think about the weight of managing the household on the cusp of winter. A foot of snow fell about a week after the tragedy. It's no small wonder that widows with older children could not function without keeping some of their sons and daughters out of school to help sustain the household.

On the afternoon of the tragedy, CCI marshaled a group of its people to begin visiting the households of all men on duty on that day to assess what was needed. The visits continued through the next day.

The 1926 Pension Department reported:

> This force consisted of Miss Welander, Miss Atkins, and Miss Hirwas, the [CCI] visiting nurses of Ishpeming, Negaunee and Gwinn; Walter E. Johnson and William Kellow of this office; J. Henry Williams of the Safety Dept.; Miss Brown of the hospital [the precursor of Bell Hospital]; C.W. Nicolson

and H.C. Moulton of the Engineering Dept.; J. L. Hyde of the Purchasing Dept.; and Joseph Pope and John Peterson of the North Lake and Barnes-Hecker offices.

The nurses, with Miss Brown, continued in their work of visiting the homes and I [William H. Moulton] personally saw all of the families very shortly after the accident.

One of the most remarkable things was the fine feeling manifested by all of the families without a single exception, and mention must be made of Mrs. William E. Hill, wife of the County Mine Inspector, who felt it almost impossible to believe that any company would help her and pay the regular workmen's compensation as the Cleveland Cliffs Iron Company were doing.

Immediately following the accident, the National Red Cross telegraphed to ascertain if help was needed and the Marquette County Red Cross officials wired that everything was being taken care of. However, they sent from St. Louis Miss Miller, Field Director, who arrived in Ishpeming on Friday, November 5th and a conference was held in this office with the following persons: Miss Edith R. Miller; K.I. Sawyer, Mayor of Ishpeming; Mrs. C.C. Cowpland, President of The Women's Services Club; R.S. Archibald, Chairman of the Marquette County Red Cross; Miss M.S. Brown of the Hospital, Miss Myrtle Welander, Miss Ina E. Atkins, and Miss Johanna L. Hirwas, our three nurses and the Secretary.

William Moulton added, "I wish to express my appreciation of the work of all of the assistants in this department who following the accident worked untiringly throughout the days, nights and Sundays on work of all kinds, both in and out of the office, which had to be done promptly and accurately."

The report further stated that the Marquette office of the Red Cross provided $500 for use by Mrs. A. L. Johnson, the county nurse for any special needs of the families. As Thanksgiving and Christmas approached, there were individuals and organizations throughout the County that provided special help to some of the families. Between the time of the accident and the beginning of benefits, CCI provided additional help with fuel, food, clothing and other basic necessities that varied depending on what the families needed.

CCI self-funded its workmen's compensation, with payments **of $14 per week for five years,** and those payments began as soon as documents were signed by the next of kin. On January 11, 1927, when all pumping operations proved futile in attempts to recover the bodies, CCI President William G. Mather declared that **CCI would provide a dollar-for-dollar match to the workmen's comp, resulting in $28 per week, with the families requesting payments every two weeks. In all, a total of $8400 was paid in installments over a period of five years.**

In some cases, the compensation was more than the *miner* earned. Some of the widows requested partial lump-sum payments, which were used to purchase homes, pay for medical expenses, and pay off debts.

Compensation for families of the unmarried men was calculated based on the proportion of support that was provided by each of the lost men.

In 1926, churches of all denominations did not conduct funerals if no remains were recovered.

The anguish of the widows of the Roman Catholic Faith was exacerbated because none of the rituals of their faith were permitted for the men whose remains were not recovered. Many widows and family members held out hope against hope that their men would be recovered. By the beginning of December, all of the widows except those in Finland had signed the paperwork needed to receive the compensation.

From time to time *The Daily Mining Journal* continued to print short articles about donations from fraternal and service organizations, and sewing groups that gathered to mend clothing. Compassion ran deep, as it still does in Marquette County, and the private interactions and acts of kindness, neighbor to neighbor, family to family, helped to sustain the widows and children through the aftermath of the worst mining disaster in Michigan History.

4 True Miners' Wives

On November 8, 1926, the *Daily Mining Journal* printed an eloquent tribute to the grieving widows, based on the reports from the nurses and social workers who had visited in the homes of the lost men. Here is that tribute verbatim:

Fathers live in children

They are rallying all their strength to take care of the small girls and boys, all that is left of the dead loved ones. Those women are finding comfort in an old and deep-rooted philosophy of life. In the boys and girls the fathers' spirit still lives and the women are bending their efforts to tending those children.

While the boys and girls are the greatest comfort the women have, they are also unconscious sources of grief. The little ones stand about wide-eyed, not quite understanding what it is all about, but sure that things are not well at home, else mother would not look sad.

A mother in one of the stricken families brought home dolls to her fatherless little girls.

Childlike they crooned their delight, then exclaimed, "We want to show daddy!"

That's the thing that hurts, but the mothers only press pale lips a bit closer and go bravely on.

When some of the social workers went to one mother, they found her bent over her sewing. Silent, heartbreaking tears slid down her white cheeks and dropped on the woolen cloth in her lap. The women wanted to know what they could do for her.

A search of the accident reports that list the names and ages of the minor children shows that Marian and Walter Tippett's twin daughters, Dawne and Dixie, appear to be the only sisters who are close enough in age to be the recipients of the dolls. Dawne's daughter Melody Smail Visser gifted the matching dolls of her Mom and Aunt Dixie to the collection of the Marquette Regional History Center. Dawne married Lowell Smail, and Dixie married Lowell's brother Hugh.

Ask not for self

She didn't ask anything for herself. That is not the way of those women. Her plea was for the family.

"I have been trying to get the clothes sewed for the children to wear to the funeral, but I don't seem to be able to hurry." Stubbornly she winked away the fast-gathering tears, and faced her visitors. "Do you think I could borrow some clothes for the children? I want them to look [nice] when they go to their father's funeral."

Abnegation of desire, forgetting of self and their plight in an effort to care for their little boys and girls, willing to accept the cross which they cannot understand, crucified a dozen times a day as they come across an old pipe, a dented dinner pail, the pan in which they baked the pasties the men liked for lunch, hundreds of homely objects speak eloquently to those women of the beloved!

Let no one be misled by the silence of their grief. The world for those women is filled with pain. Only they know of the hours when they lie, at night, staring straight into the darkness with unseeing eyes, thinking, thinking, but never of themselves, always of those who have been snatched away from them, and of those who remain.

Their fortitude may impress an outsider, but it does not surprise those who know them well. Those women are brave souls; they are true miners' wives.

"Brave souls" – some of the widows and children of the lost men. Back row, left to right, Nanna Santti, Vieno Koskinen, Aili Franti (Lehtimaki), Miriam Laituri (Johnson), Ellen Luoma, unknown, Anne Kirby. Front row, Anna Franti, Ruth Laituri (St. Martin), Alle Laituri, Sanna Kallio holding baby Taimi (Walimaa), and Toivo Kallio. (Courtesy Michigan Iron Industry Museum and Marquette Regional History Center.)

5   Men Remaining on Day Shift

On the day of the catastrophe, who was left on the surface after the cave-in?

With 50 men lost underground, there were as many as 20 left on the surface. The list below was compiled from documented sources such as the coroner's inquest, CCI records, and the October 31, 1926 payroll. If there are no comments with the names, nothing is known about whether or not they were on duty that day.

> Allyvion Miners, undergound track foreman, came to the surface a little after 11 am. (See Chapter 10: "It seemed as though the *shaft* was tore apart.")

> Santoke Combelli, underground track helper, accompanied Al Miners on the cage that came up a little after 11 am.

> Andrew Malvasio, cage rider, was mentioned in Allyvion Miners's inquest testimony.

> Ed Hillman, pipe foreman, was riding the cage to the surface when a strong wind was felt in the shaft around 11:20 am. (See Chapter 9: "A big wind in the shaft.")

> Albert Tippett, pipe man, came to the surface with Ed Hillman. As the accident began to unfold, Albert followed Hillman down the ladder into the Barnes Hecker shaft. Al's daughter Patricia Procunier said something fell from above and struck him on the shoulder. It left a lump that remained for the rest of his life.

Arthur Francis, blacksmith and/or Joseph Peppin [Pepin], blacksmith helper/assistant surface foreman. Al Miners said he was in the blacksmith shop after he arrived on the surface. Timber yard worker T. Wilfred Tippett said Joe Pepin was the assistant surface foremen.

John T. (Jack) Wills, dryman and pump helper. Jack was on duty when his son, Rutherford Wills, arrived in the first-aid station in the "dry" building after his 800-foot climb to safety. Jack was also step-dad to William, Walter, and Albert Tippett. (See Chapter 12: "Lone Survivor")

J. A. Peterson, mine clerk, whose signature is on almost all payroll records and the accident reports. (See Chapter 6: "Keeper of Names," and Chapter 3: "Help for the families")

Joseph Pope, mine clerk, is named in the 1924 annual report as an additional clerk at Barnes-Hecker. This timing coincides with the addition of the second shift in July of 1924. In "Help for the families," based on 1926 records, Pope is named as a clerk at Morris-Lloyd.

T. Wilfred Tippett, surface laborer in the timber yard. "Tip" was about 18 when he worked in the timber yard at Barnes-Hecker. During his career in mining, Tip become mine captain of the Lloyd and Mather B mines. His father, Thomas Tippett, was captain of the Lloyd Mine at the time of the Barnes-Hecker tragedy. (See Chapter 11: "Timber")

Richard Lucas, hoist operator. (See Chapter 8: "In that moment")

Roy Poirier and/or Joseph Hebert, timber *trammers*. Their job was to load and push carts filled with logs and timbers used for lining the tunnels and *raises*. T. Wilfred Tippett cites Roy Poirier as his partner in the timber yard on the day of the tragedy. (See Chapter 11: "Timber")

Charles Dellangelo, electric foreman, mentioned in recollections of T. Wilfred Tippett.

Horace Kirschner, shovel foreman

William Westermeyer, shovel operator

Ed Billings, John Lawry, Edlore St. Andre, and Louis Trewella, shovel laborers

Zephire Nault and Nickolai Lahti, carpenters

6   Keeper of Names
J. A. Peterson (1894-1957)
John Algot Peterson, mine clerk

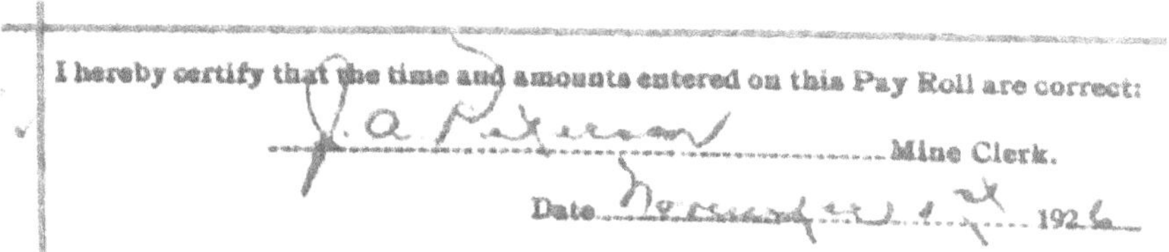

He is a man known only by his signature, which appears on almost every page of every Barnes-Hecker payroll ledger from 1921-1927. J.A. Peterson was the mine clerk.

When mayhem broke loose on the surface of the Barnes-Hecker Mine on the morning of November 3, 1926, the mine office would have become a hub for communication, and it may have had only one phone connected to regular phone lines.

What was it like dialing the Morris-Lloyd to reach North Lake District Superintendent Charles Stakel? The CCI main office? Unknown numbers of other offices and individuals? And handling incalculable numbers of incoming calls.

As the sobering truth of death became apparent, Peterson would have been involved in the grim task of verifying the names of the lost and assembling a list for the CCI main office. Thankfully, a second person, Joseph Pope, also worked in the mine office.

In *The Daily Mining Journal* EXTRA that hit the streets at four o'clock in the afternoon on November 3, the initial number of fatalities was believed to be as high as 65. The paper reported that there were between 40 and 65 men trapped underground. The actual number was corrected to 51 when the names were released. Mr. Peterson knew these men personally. He knew those who were left on surface and the other men who weren't on duty.

There were 51 accident reports to be prepared, one for each of the lost mine employees, and one for Rutherford Wills, who had escaped the flooding mine. Although the accident reports are all dated November 3, 1926, they weren't all signed until early December.

Census records for 1920 and 1930 show John A. Peterson as mining clerk and bookkeeper in an iron mine. By matching census records that include his wife Rose, we found that John Algot Peterson and Rose Peterson are buried in the same plot in the Ishpeming cemetery.

The narrative of the 1926 annual report for the CCI pension department shows that John Peterson is, indeed, named as the mine clerk for Barnes-Hecker. He was one of the CCI officials who visited the homes of all the employees who were at work at the mine on November 3.

Other records led to the descendants of J.A. Peterson, and they now have a missing piece of their family history.

In a 1918 series of professional photographs taken at the Barnes-Hecker Mine site, there is one photo that could plausibly be J.A. Peterson.

Barnes-Hecker Mine view showing the office and engine house, September 1918. Although the man in the photo is not identified, his attire suggests that he could be mine clerk J.A. Peterson.

That can never be verified, but the photo is presented here as a means of honoring a man who must surely have suffered a heavy emotional burden by being the keeper of the names of the lost men, and those who toiled with them at Barnes-Hecker. (See also: "Men on surface;" "Help for the families")

Barnes-Hecker payroll ledgers, 1924-1926; Cleveland Cliffs Iron Company (Courtesy Central Upper Peninsula and Northern Michigan University Archives).

7

5 A.M.
John Glanville (1889-1975)
Night shift boss, recovery shift boss

At 5 am on Tuesday, November 2, 1926, Night Shift Boss John Glanville finished his nightly rounds through the workings of Barnes-Hecker and took the cage to the surface. The night shift crew finished up at 8 am, and because November 2 was Election Day, the mine was closed until the day shift on Wednesday, November 3.

Glanville would later testify at the Coroner's Inquest that he had visited all areas of the mine twice on his nightly rounds, and found all in order. He said the only person expected to be in the mine until the morning of November 3 was a pump man.

Glanville would later become one of the men assigned to supervise round-the-clock pumping and recovery efforts in the wake of the November 3 disaster. Others included Al Miners and Charles Pepin, both of whom had also worked as shift bosses at the mine. Born in 1888 in Cornwall, England, Glanville arrived in Ishpeming when he was 18. By 1914, he was a 26-year-old miner at Cliffs Shaft Mine. At that time, he participated in an 18-month, voluntary curriculum of courses offered by CCI for men interested in becoming shift bosses. By the end of his career, he had worked 45 years for CCI, retiring as mine captain at Cliffs Shaft. He was 86 when he died.

John Glanville [Person #23 cropped from CCI photo of men in 1914 Education Department class] (Courtesy of Central Upper Peninsula and Northern Michigan University Archives)

8 In that Moment
Richard and Helen Lucas
Hoist Operator

On the morning of November 3, the students at Barnes-Hecker School had one eye on the clock as the lunch break drew near. Some had to run home and carry their fathers' lunch pails to the mine.

Nine-year-old Richard Lucas was one of those kids. He loved seeing his dad run the giant hoisting machines in the engine house. On this day, his dad was intent on his work and said, "you can put the lunch down over there, son, and go on home."

Barnes-Hecker School students in 1924. The boy in the first row, far right, is Richard Lucas, who was nine years old when the mine caved in. The descendants of Peter Mongiat identified the two girls in the second row, second and third from left as Mary and Sena Mongiat. (Courtesy of Tom Lucas)

Helen Lucas with son Richard in Barnes-Hecker Location, ca. 1918-19, summer and winter. (Courtesy of Tom Lucas)

The Lucas family lived in the Barnes-Hecker Location, in the first duplex next to Mine Captain Bill Tippett. As Richard came up the path on his way home, Mrs. Tippett intercepted him. "Richard, what's going on at the mine? The power's out," she said.

In that moment, the mine siren sliced through the air, and they knew there was trouble at the mine. In that moment, Nellie Tippett became one of 41 widows. And in that moment, the boy became the last known person to go to and from the mine before disaster struck.

In 1918, when the new homes in the Barnes-Hecker Location were completed, the Lucas family was one of the first to move in. Mr. Lucas's employment record shows he started working at the new mine on September 1, 1918 as a motor brakeman. At that time, the shaft had been sunk and the main *levels* were in progress.

In the hours, days, weeks and months following the moment when disaster struck, the senior Richard Lucas went to work every day as a surface laborer. He was there when remains of co-workers Jack Hanna,

THE CLEVELAND-CLIFFS IRON COMPANY — Employment Record

782

12-21-17.

MAY 15 1928

Date of Birth August 9th, 1889 Age 28.	
Place of Birth	RICHARD
Nationality American	
How Long in U. S.	
U. S. Citizen Yes First Papers	
Married Yes Single No. of Children 1	
Ages of Boys 2 mos. Ages of Girls	
Reads Writes And Speaks English	
Street and Number City With Whom Living	
North Lake Ishpeming	

RICHARD   IUCAS

FIRST NAME   INITIAL   LAST NAME

RECORD No. 2202.

DATE EMPLOYED	MINE	OCCUPATION	BRASS CHECK NO.	DATE LEFT	REASON
12-13-17	Morris-Lloyd	Pufferman	254	8-15-18	
9-1-18	Barnes-Hecker	Motor Brakeman	5	4-19 27	trans
4-15-27	Morris Lloyd		19 9	4/25/27	quit

Richard Lucas employment card (Courtesy of Tom Lucas)

Joe Mankee, and Thomas Kirby, Jr. were brought to the surface. As time passed, the mine continued to ship ore from the stockpiles, and equipment including the giant hoisting engines was removed.

At the end of May 1927, Lucas accepted a job at the neighboring Morris Mine. After two weeks, he resigned, packed up his wife and son and whatever belongings they could carry in their truck, and left for Iron Mountain, where he

Richard Lucas retirement portrait (Courtesy of Tom Lucas)

took a job in the Ford factory. Over time, Helen LaFond Lucas found it harder and harder to be away from family, so the Lucases returned to the Ishpeming area where Mr. Lucas worked at the Lloyd Mine in the North Lake District until he retired.

There was no consolation for the men who were not lost in the Barnes-Hecker tragedy. Tom Lucas said his grandfather had seen too much, and Mr. Lucas lived with the shadow of PTSD for the rest of his life. But he didn't give up. Tom Lucas described his grandfather's legacy with two words: hard work.

Mr. Lucas once built a rock wall in his yard, and it remains to this day. Perhaps the moment-by-moment placement of those stones illustrates perseverance. He did what he had to do for his family, moment by moment, until the moments added up to a lifetime.

– Tom Lucas

9   A Big Wind in the Shaft
Allyvion Miners (1892-1940)
Underground track foreman

Allyvion Miners was hired at the Barnes-Hecker in 1917 after the shaft was completed. According to his 1940 obituary, he was the *drifting* boss, which meant he would have supervised some, if not all, of the drifting [tunneling] operations to open the three main levels at the mine. By the time of the first ore production in 1922, he was the shift boss. The mine operated with only the day shift until 1924.

In 1926, Miners was the underground track foreman. On November 3, he had spent the morning working on all three levels before taking the cage to the surface shortly after 11 am with track helper Santoke Combelli and cage rider Andrew Malvasio.

Miners was in the blacksmith shop about 50 feet from the shaft when he heard "something like a big wind in the shaft" and went to see what was going on. At first, he assumed the wind was coming from the main air discharge or a broken airline. He saw Pipe Foreman Ed Hillman and his helper Al Tippett go down the ladder, and when he heard a second crashing sound, he started down the shaft. He met Hillman, Tippett and Rutherford Wills coming up.

According to payroll records, Miners and Ed Hillman were the only foremen left on the day shift. They may have directed the mine office to summon North Lake District Superintendent Charles Stakel, who was underground at the Morris at the time of the cave-in. In the chapter "Timber," T. Wilfred Tippett, who worked in the timber yard, said he assisted Miners in dropping a plumb line down through the bottom of the cage and found the water 45 feet from the surface.

Following the tragedy, Miners became one of a small group of shift bosses assigned to supervise round-the-clock pumping efforts that ultimately proved to be futile. On November 20, with water rising eight feet per hour, officials ordered everyone out of the shaft. About 20 minutes later, the shaft flooded.

At the coroner's inquest in February of 1927, Miners provided detailed descriptions of the "caving system" used at the mine and the amount of timber used to fill mined-out spaces before those voids were blasted down. He said he had never seen so much matting [timbers and logs] used in all his years in mining. He emphasized that there was a constant emphasis on keeping the workplace safe.

Allyvion Miners was born in Salisbury Location south of Ishpeming in 1892. His parents moved the family to Cornwall, England, when he was two years old. He began working in Cornwall tin mines when he was 14. At 23, he returned to America and lived in Arizona for a short time before returning to Ishpeming. Following the Barnes-Hecker tragedy, he became a foreman at the Morris Mine, and in 1934 became captain of the Blueberry Mine. In his lifetime, he worked in tin, copper and iron ore mines. He was survived by his wife Signe, two daughters and two sons.

His obituary stated, "He was rated as one of the outstanding mining men of the district."

Fight to Recover

BANDON ALL
OPE FOR 51
BURIED MEN

(Photo on Picture Page.)

HPEMING, Mich., Nov. 5.—
e is no visible sign of water
9 feet in the Barnes-Hecker
, rescue workers who reached
level reported late today.

:overy of the fifty-one men, be
, dead, trapped at the 1,060-foot
when a hidden lake engulfed
on mine, will be hastened if the
· has disappeared, it was said

NDON HOPE.

ding mine authorities declare
after a conference there wa
st one chance in ten millio
the surging flood which e
i the mine had not killed the
fifty-one.

· peculiar action of the wate
at first swept through th
levels and up to within 1
f the surface after the cave-i
to recede well below the 40
evel, was the slender ray
which urged on the resci
rs.

first of seven funerals f
s whose bodies already ha
recovered will be held toni
vhen William Holt will
He left a wife and sev
n.

ENT COSTLY.

lly large families have be
other victims. That of B
hapman mourns not only te
d and father, but also H·
le eldest of ten children.
of the families will recei
veek for 300 weeks, requir
company by state law. Mce
,000,000 will be expended y
mpany in recovering ie
t is estimated

Beat Death by Inches!

SOLE SURVIVOR—Wilford Mills, 23, left, the sole survivor of the Ishpeming mine tragedy, telling Al Miners, another miner, of his 800-foot climb to the surface as the water rushed through the mine. The water lapped his boots as he ascended. Two companions behind him failed to reach the top.   Herald and Examiner phot

In this photo from the Chicago Herald and Examiner, Al Miners at right is dressed in his oil coat for this posed photo with Rutherford Wills, the lone survivor. According to the Newberry Library in Chicago, the newspaper has not been published for many years, and there are no archives of original photos. (Courtesy of Tippett descendants)

10   It seemed as though the shaft was tore apart
Edward J. Hillman (1890-1941)
Pipe foreman

On the morning of November 3, 1926, pipe foreman Ed Hillman spent three hours on the third level of the Barnes-Hecker before stepping onto the cage with pipe helper Albert Tippett. They had work to do on the first level, and as the cage rose up the shaft, Hillman asked cage driver Andrew Malvasio what time it was. Twenty minutes past eleven. . . not enough time to start a large job so they continued to the surface.

Hillman felt a rush of air and assumed it came from above. At the top of the shaft, he heard two raps on a pipe—the signal asking for the cage on third level. Nothing unusual.

A few minutes later, he was in the shop when he was summoned to check for a broken air pipe in the shaft. Emerging from the shop, he heard a roaring sound. He shut off the air valve, but the roaring continued. He checked the water columns coming up from the pump house below, but found no water coming up. He called the pump house on third level. No reply. He called second level. No reply. First level. No reply.

He and Al Tippett put on oil coats and went down for a look. Power to the cage was out so they started down the ladderway. Hilllman estimated they were about 300 feet down when they met Rutherford Wills coming up. Breathlessly, Wills said three others were behind him. Hillman shined a light down but saw no one. He called down and got no response.

In Hillman's February 2, 1927, testimony at the coroner's inquest, he said, "I could feel a heavy pressure of air coming and I thought it was time to get away. We climbed up one ladder about 75 feet long and we started up the second one when the second crash came. It seemed as though the shaft was tore apart; that was the feeling I had anyway."

The three men met Al Miners coming down, and they followed his light as they sped for the surface. At the top, Hillman and Tippett reached down to grab Wills by the shirt and hauled him up out of the shaft. Wills collapsed, exhausted. They sent him to the first-aid station in the *"dry building,"* where Wills's father Jack Wills was the *dry man* in charge of the first-aid station.

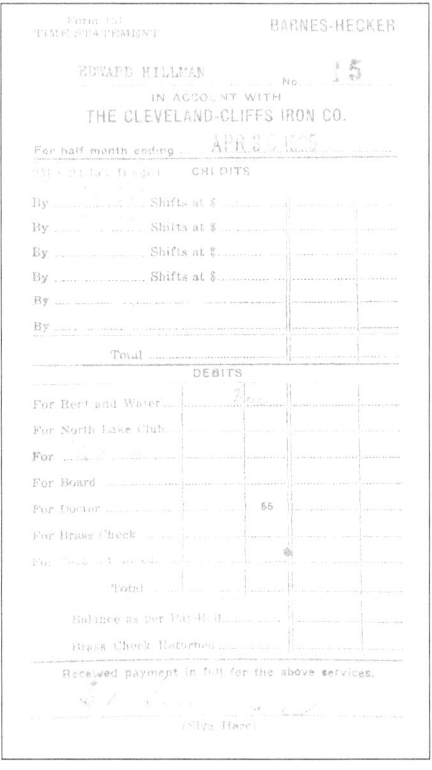

With the power and phones out, and Wills's description of water chasing him up the ladder, Hillman and Miners knew that both Mine Captain Bill Tippett and Shift Boss Sam Phillippi were cut off somewhere underground. A normal morning had flipped to a crisis in the span of 10 minutes.

Hillman and Miners were the only men in the payroll records who were identified as supervisors. They may have given the order to summon North Lake District Superintendent Charles Stakel, and briefed him on what was known.

In the days that followed, Hillman was one of the men assigned as shift bosses overseeing round-the-clock pumping in the attempt to dewater the mine. The effort became too dangerous, and the recovery crews in the shaft were ordered out on November 20. Pumping efforts continued until January 11, when recovery was deemed impossible. On

January 12, the banner headline in *The Daily Mining Journal* read: "CCI to abandon Barnes-Hecker."

Hillman must have been keenly aware of his wife Louise, who was eight months pregnant with their sixth child. What would have happened to his family if he'd been swallowed up by the rising water?

Ed Hillman had worked at the Barnes-Hecker since the beginning in 1917. By the time of the tragedy, he had worked in mines 12-14 years. He continued to work at the Morris-Lloyd/North Lake Mine until August 1941, when he succumbed after a debilitating illness. He was survived by his wife Louise, three daughters and three sons. Hillman's family gifted one of his time cards dated April 20, 1925, to the Michigan Iron Industry Museum, where it is on display.

<table><tr><td>

11

</td><td>

Timber
T. Wilfred Tippett, timber yard laborer (age 18)
Captain, Mather B Mine

</td></tr></table>

Thomas Wilfred "Tip" Tippett, who served as mine captain of several mines on the Marquette Iron Range, was an 18-year-old worker in the timber yard when catastrophe struck at the Barnes-Hecker. In 1989, he wrote recollections about what he experienced that day:

> Wednesday morning, November 3, 1926, was cool with a little bit of snow on the ground, but the men waiting for the 7:30 a.m. cage to take them underground, were dressed for the cool temperatures in the mine. I, T. Wilfred Tippett and my partner Roy Poirier were part of the crowd and having a great time joking and discussing the voting on the day before. Tuesday was a national election day and a holiday for the people who worked at the mines. It was a special occasion for me because Uncle Walter Tippett was starting his first day of work at the Barnes-Hecker mine. He was a real pal of mine because we were very sports minded and also participated in football and baseball. Plus Uncle Walter was very well-known throughout the Upper Peninsula and Wisconsin as a great catch-as-catch-can wrestler. I tried to get his clothes dirty, but I failed.
>
> After all the men went underground, Roy and I got a supply of empty underground trucks from the mine, and we proceeded to load them with supplies needed to mine the iron ore. These supplies consisted of hardwood timbers 9 ft. long, tamarack poles 9 ½ feet long, split cedar lagging 5 ft. long and tamarack cribbing 5 ft. 4 in. long. These supplies were used to support the drifts and raises.

Roy and I worked all morning loading the trucks and had pulled them from the timber field to a spot near the shaft house to be loaded on the cage to go underground. The time was 11:00 a.m. and we would have to wait until after lunch hour to load the trucks onto the cage because it was near the time for the men working underground to come to the surface for their lunch.

At 11:40 a.m. the first concussion of the cave in occurred. It blew dirt and moisture from the mine at a terrific pressure. The cage in the shaft was on its way to the surface but the electricity was shorted out at the main switch and the hoist pulling up the cage was only about 15 feet from surface. Charles Delangelo [sic], the electric foreman was in the electric shop near the shaft house. He ran to the switch near the hoist building and got the electricity back so the hoist could operate and bring the cage to the surface. The men in the cage as I remember were Andrew Malvasio (the cage attendant or rider), Ed Hillman and Uncle Albert Tippett.

The rumbling in the shaft was very loud with a great gust of wind being pushed up the shaft. No one knew what had happened and of course no one knew what to do. When the first concussion came, the top tram car (which takes the iron ore when the skip empties into it, out on a trestle and dumps it onto a pile for storage), was stalled out on the trestle. The rope had jumped off one of the guides. The assistant surface foreman Joe Pepin had picked my partner, Roy and I to help put the rope back where it belonged so the tram car could be pulled back to the shaft house. While Joe was on the landing which is about 20 ft. high above the ground, he noticed a hole in the ground about a quarter of a mile from the mine shaft, so he and someone went to investigate the hole.

I do not remember the exact time that Wilfred [Rutherford] Wills got to surface after his climb to safety. I know that Uncle Albert went down the shaft on a ladder when Wilfred hollered for help. One cannot help a person climbing a ladder very much but moral support is a great thing when one is in trouble. When Wilfred Wills (whose given name was Rutherford) climbed onto surface, he fainted from sheer exhaustion. He recovered and was o.k. after a few minutes. He walked over to the dry house where the locker and water showers were. [Wills's father, Jack Wills, was the dry man, and the first aid station was located in the dry.]

I was with Al Miners when we went into the cage to drop a carpenter's string through the hole in the floor of the cage to measure the water in the shaft. That was 45 minutes after the first concussion and the water was 45 feet from the surface. It receded a bit afterward but I know how close it got to the surface.

When Joe Pepin got back to the shaft house after investigating the hole in the ground we were all told to get shovels and picks and to proceed to the hole in the ground to stop the water from pouring into the hole from an underground stream. Of course after trying to do so, it was just too dangerous to continue. The ground all around the edges was still caving into the big hole that had caused the mine to fill up with water and debris. There was real anxiety by now for all of the men working underground, but no one knew what to do to help the situation. When we got back to the shaft house Mr. Stakel the Superintendent and Pete Allen the Mining Engineer were on the premises and had taken charge. It was a real relief for everyone.

The orders were given to get pumps to the cave in area and electricity and the water bailers were on the way from the Negaunee area to be used in the shaft to lower the water. These bailers are used like a skip, they are attached to the skip steel rope and lowered into the water until they are filled, then hoisted to the surface and emptied. The work to get one new rope on the drum in the engine house, where the hoisting engines are and to get the skip out of the shaft was started.

By now word had spread throughout Marquette County of the Barnes Hecker cave in and the people started to come out to the mine area, causing quite a problem for the workers, so a temporary rope barrier had to be put up to keep the crowd away from the work area. The people also wanted to be helpful and some of the young men were used to keep the crowd back of the rope area.

Every available piece of machinery and materials from other mines was sent to Barnes Hecker via trucks and the work proceeded very well. I am not too sure what time the first bailer full of water came to surface, but I think it was near midnight. I know it was early morning by the time I got to my home in North Lake. It was evident that all the men were trapped in the mine.

Our family had lost, Uncle Bill Tippett the captain of the mine, and Uncle Walter Tippett, his first day of work at the mine. Uncle Albert Tippett was lucky to be alive, because he had talked to Uncle Bill Tippett on the 3rd level 1,000 feet below surface, before he entered the cage to come to surface. He wanted to know if Uncle Bill and Hill the County Mine inspector were coming to the surface with Albert, Hill and Andrew Malvasio. Uncle Bill said he and Hill were to inspect one more place and that would complete Hill's mine inspection. The cave in occurred before the cage reached the surface, (but only about 15 ft.) when the electricity was cut off.

When I reported to work on Thursday the surface crew were detailed to clean up around the shaft and helping to install another water pump at the surface of the cave in. The pumps had to be put on rails because the cave in hole kept getting bigger and the water level had not decreased at all. I can still see the underground stream pouring into that hole.

I think I was lucky to be looking at the cave in instead of in it because the previous Sunday we hunted rabbits in the area. The hunting party included my Dad Tom Tippett, Uncle Walter Tippett, Uncle Bill Tippett, Wilfred Wills, Uncle Albert Tippett, Al Miners, Tom Drew and myself hunted snow shoe rabbits in the area of the cave-in, because it was swampy land and lots of rabbits were in that area.

I don't know how many men who worked at the Barnes-Hecker are still alive. My statements are from the memory of that day.

T. Wilfred Tippett
May 23, 1989

According to his obituary, "Tip" was born in 1909 in Calumet, Michigan, a son of Thomas and Bertha Tippett. By the time of his passing, he had worked in mining for 45 years. Following the Barnes-Hecker disaster, he worked at the Morris-Lloyd and the Negaunee Mine. He served as mine captain at the Negaunee, Cambria, Jackson, Athens and Mather B Mines.

His community involvement included Boy Scouts, the Ishpeming Ski Club and a Kiwanis Club in Arizona. He was a member of the Ishpeming United Presbyterian Church for more than 50 years. A sports enthusiast, he played semipro football in his youth and golf until two years before he died. In retirement, he and his wife Lyle lived in Maricopa County Arizona, where he died January 29, 1999. He is buried in the Ishpeming cemetery.

He was predeceased by his first wife Mina, his parents, brothers Douglas, Roy, and Leonard, and survived by a daughter and several grandchildren and great-grandchildren.

See also: Chapter 20: Catalyst for change; Chapter 26: A grateful heart, Chapter 41: Legacy bearers

12 Lone Survivor
Rutherford "Wilfred" Wills (Motorman, 22) and Bruna Phillippi Wills

Wedding of Rutherford Wills and Bruna Phillippi Wills in September of 1926. Attendants were Margaret Tippett, daughter of the Mine Captain Bill Tippett, and Joe Mankee, who perished in the Barnes-Hecker tragedy. (Courtesy of Tippett descendants)

In September of 1926, Rutherford Wills married Bruna Phillippi, whose father, Sam (Sante Filippi), was the day shift boss at the Barnes-Hecker Mine. The newlyweds lived with Rutherford's parents, Mary and Jack Wills, in one of the duplexes in the Barnes-Hecker Location.

On November 3, 1926, Rutherford was a motor man on the second level of the Barnes-Hecker, where he was working with his brake man, Jack Hanna. They were 800 feet below the surface.

Shortly before 11:20 in the morning, as Wills was heading away from the shaft to pick up filled ore cars, he felt a gust of air and stopped his motor 400 feet from the shaft. He told Jack, "We will stop here. Something's wrong." The power went out. A second rush of air blew out the men's carbide lamps, and Wills said, "We better get out of here."

Rutherford Wills: lone survivor to escape the flooding B-H shaft

The two men headed for the shaft, kicking their boots against the railroad tracks to find their way.

At the shaft, they urged *skip tenders* Joe Mankee on second level and Tom Kirby Jr. on first level to follow them up the ladder. Wills was lanky and lean, and he led the way. He estimated that about 75 feet above first level [i.e. 525 feet from the surface] he saw light coming down from carbide lamps and encountered his half-brother Albert Tippett and others climbing down. "Save yourselves," he said. As he reached the surface, he felt the grip of his brother and another man hauling him up to safety, where he lay gasping on the ground. The

800-foot climb had taken 10 minutes. He had climbed the equivalent of 2/3 of the height of the Eiffel Tower.

Wills was taken to the shelter of the Dry, where his father, dry man Jack Wills, took charge. When Bruna arrived at the Dry with lunches, her father-in-law had to turn her away. "Go home," Jack said, "Your husband's out. He's okay." Wills's accident report would list his injury as "Nervous shock," now known as PTSD.

Rutherford was the only man to escape the flooding Barnes-Hecker Mine. His half-brother Albert had gone to the surface shortly after 11am, and his half-brothers Bill and Walter Tippett were among the dead. Rutherford and Bruna also mourned the loss of her father, Shift Boss Sam Phillippi, and their best man, Joe Mankee.

Everywhere he went, people wanted to hear the story of Wills's harrowing climb, and reporters were eager to have his story. In one of his interviews with the local newspaper, he said, "You can tell the world for me, that I am finished with mining."

Rutherford and Bruna Wills in the Barnes-Hecker Location shortly after the tragedy. (Courtesy of Tippett descendants)

As time passed, hope sprang up when Rutherford and Bruna learned that they were expecting a child in the spring. After their son was born, they moved to Flint, where Rutherford accepted a job as a security guard in the auto industry.

In 1927, the CCI Pension Department reported that after the mine tragedy, doctors evaluated Wills, saying "he became extremely nervous… his condition similar to the shell shock cases of the war." As a result, Wills received $1,372 in compensation.

In 1942-43, Rutherford's mother, Mary Ellen Tippett Wills, was in failing health in North Lake. Rutherford and Bruna returned to North Lake for about a year during that same time, and Rutherford accepted a surface job at the Mather A Mine in Ishpeming. Presumably, after Mary Wills died in March of 1943, Rutherford and Bruna returned to Flint, and he noted that he would never work in mining again.

For decades, he suffered nightmares and woke up in a cold sweat, clutching the drenched bed sheets. "I must have climbed out of that mine 100 times in my sleep," he said.

In 1972, Rutherford was in the late stages of cancer. With special consent from his doctor, he traveled to Ishpeming in August to attend the dedication of the granite Barnes-Hecker monument commissioned by the citizens of Ely Township to commemorate the worst mine disaster in Michigan history.

Although his name does not appear on the monument, Rutherford Wills's name is etched into the history of the Barnes-Hecker tragedy.

(See also: Chapter 40: Ely Township monument and Chapter 41: Legacy bearers; Chapter 53: Man's impermanence)

13

Best Man
Joseph Mankee (22)
Skip tender

Joe Mankee depicted at Wills' wedding

Joe Mankee had been the best man at the wedding of Bruna Phillippi and Rutherford Wills, two months before the Barnes-Hecker Mine tragedy. He was engaged to Margaret Tippett, daughter of Mine Captain Bill Tippett and his wife Nellie.

Mankee immigrated from England in 1922, and began working at Barnes-Hecker in March of 1923. He lived in the Barnes-Hecker Location with his mother Elizabeth Pope and step-dad Edwin Pope, who also worked at Barnes-Hecker.

On the morning of November 3, 1926, Mankee was a skip tender on the second level when the power failed and gusts of air blew out the carbide lamps on the men's helmets. Mankee was one of the three men who raced up the ladderway behind Wills, but were caught in the melee of water, mud and debris that took their lives. His body was recovered from the shaft on November 8.

Joe Mankee was survived by his mother Elizabeth Pope and stepdad Edwin Pope, brothers Thomas, Llewellyn ("Buster"), and Jack Pope still living at home, two sisters in England and two sisters, Olga and unknown in the United States. He was a member of the Sons of St. George. Because Mankee was unmarried, his mother, Elizabeth Pope, was the recipient of the death benefit paid by CCI.

14 Lanterns in the Dark
Charles J. Stakel (1883-1982)
North Lake District Superintendent

On the morning of November 3, North Lake Mine District Superintendent Charles Stakel had promised his wife the use of the family car, so instead of making rounds at the Barnes-Hecker, he walked to the nearby Morris-Lloyd to make rounds there.

According to his inquest testimony, Stakel was unaware of trouble at the Barnes-Hecker until he took the cage from the seventh level of the Morris to the surface. He was told that Ed Hillman had called from the Barnes-Hecker to say there was something wrong. When Stakel returned the call, he was only told there was some sort of accident underground. Stakel changed out of his mining clothes and drove to the Barnes-Hecker site.

When he saw the disastrous situation, one of his first concerns was for the safety of the men in the neighboring Morris Mine, which was connected underground to the flooded Barnes-Hecker. The Morris was evacuated.

When Stakel returned to the Morris, he and day shift boss Ernest Carlyon ventured into the 5,300-foot connecting drift to check the integrity of a concrete *bulkhead*. Records make it possible to reconstruct what it was like as they made their trek into the unknown.

In the darkness of the drift, Stakel and Carlyon had only their carbide lamps and lanterns to light their way. They came upon sand that became deeper and deeper as they approached the bulkhead located about 3,000 feet from the Morris shaft. A thousand feet farther along, they were 1,000 feet short of reaching the Barnes-Hecker escape ladder. As their lamps pierced the darkness, they saw a four-foot-deep tangle of logs, mud, water and debris.

Charles and Charlotte Stakel are seated with daughters T.J. (Charlotte) Gulyash, Esther Barbara Stakel Clancey, Gretchen Stakel Seielstad and Frances Stakel Nelson seated, left to right, behind them. (Courtesy of Jay Clancey)

Their lamplight shone on what they could not have anticipated... the lifeless bodies of Bill Tippett, Mine Inspector William Hill, and Tom Kirby Sr. partly submerged in the water. Stakel had phoned Hill that morning to ask that Hill fill in for him on a routine inspection of the Barnes-Hecker. Farther back in the tangle were the bodies of four more men, most of whom were unrecognizable. Nothing could have prepared Stakel and Carlyon for what they saw. Carlyon resigned four days later.

The CCI officials knew that the connecting drift between the Barnes-Hecker and Morris sloped downward until it was 75 feet lower on the Morris end. If the mud and debris had not plugged itself in the Barnes-Hecker escape passageway, and the bulkheads had not stopped the inundation from going farther into the connecting drift, the men in the Morris Mine would have been in jeopardy. Stakel himself could have been killed.

In the early afternoon of November third, bailers were hung in the Morris shaft. Three wooden bulkheads were constructed in the connecting drift, and after sand was cleared away, Stakel oversaw construction of a 20-foot concrete plug near the existing concrete bulkhead. After the safety of the Morris was deemed secure, the Morris men were allowed to return to work. In the meantime, the Morris men were put to work in the neighboring Lloyd Mine.

Charles Stakel had become superintendent of the North Lake District in March of 1925, and typically made weekly rounds at the Morris-Lloyd and Barnes-Hecker. His last comprehensive rounds in the Barnes-Hecker were the last Friday in October… four days before the tragedy.

By the time of the Barnes-Hecker tragedy, Stakel had worked in mining for 21 years. He earned his mining engineering degree from the Michigan College of Mines, which is now Michigan Tech University. He said that prior to his work in the North Lake District, he had been in almost all CCI mines in the past 20 years.

He was assistant engineer at the Ashland Mine in 1906; from 1906-1921, he did significant survey work for the hydroelectric dam systems in Marquette and Alger Counties, then worked on the third shaft of the Negaunee Mine before becoming superintendent of the Republic Mine in 1916. He became superintendent of the North Lake District on March 15, 1925. In 1929, his responsibilities expanded to include supervision of mines in Ishpeming along with other duties. In 1943, he succeeded S.R. Elliott as mining manager for CCI. He retired in 1950.

Stakel served as superintendent of Republic Township, secretary of the Lake Superior Mining Institute, president of the YMCA, supervisor of Ishpeming Township, chair of the finance committee of the Northern Michigan Diocese of the Episcopal Church, and chairman of the Marquette County Road Commission.

Charles J. Stakel was born in Menominee, MI in 1883, and on December 2, 1916, he wed Charlotte Elizabeth Nelson. They had four daughters, Gretchen Stakel Seielstad, Frances Stakel Nelson, (T.J.) Charlotte Stakel Gulyash, and Esther Barbara Stakel Clancey; and six grandchildren.

His descendants said he was a dedicated family man. Grandson Jay Clancey said his grandfather was fondly known as "Smilin' Charlie," and he was so physically fit that he swam laps in Lake Michigamme until he was in his 80s. From 1970-71, he recorded 23 tapes of his memoirs. He died in 1982.

Stakel's far-reaching legacy is chronicled in the book, *Memoirs of Charles J. Stakel, Cleveland-Cliffs Mining Engineer, Mine Superintendent, and Mining Manager,* available from the gift shops of the Marquette Regional History Center, the Michigan Iron Industry Museum, Cliffs Shaft Mine Museum, and on Amazon.

– Jay Clancey

15 I would have made my bed there
William Conibear (1877-1956)
Safety inspector, CCI

William Conibear was at the Republic Mine in Republic, Michigan, on November 3, 1926, when he received a phone call at 12:30 in the afternoon. By then it was known that no one could be left alive at the Barnes-Hecker.

When Conibear arrived at the mine, he assessed the situation and went to find North Lake District Superintendent Charles Stakel at the nearby Morris-Lloyd Mine. There, he learned that Stakel and another man had gone into the underground drift connecting the Morris to the Barnes-Hecker and discovered the bodies of seven men in the blocked passageway.

The Daily Mining Journal identified Stakel's companion as Morris Mine Shift Boss Ernest Carlyon, who resigned on Saturday, November 6, and left for Flint. The paper reported that Carlyon vowed he'd never work in a mine again.

William Conibear was appointed CCI director of safety in 1911. He retired as safety superintendent in 1942.
(Courtesy of Michigan Iron Industry Museum)

It was 4 o'clock on November 3 when Conibear and four bosses from the Morris-Lloyd went into the connecting drift to retrieve the bodies located about 1000 feet from the connecting raise between Barnes-Hecker and Morris. The drift was almost a mile and a half long, and they took a motor and rail truck as far as they could.

Conibear said the track between the two mines had not been used by a motor, and they had to stop to clean the track switches as they went.

Conibear and Captain Nault from the Morris went on foot for the last 300 feet, making their way through a four-foot deep "soup" of water, working their way over obstacles of tangled timber and logs, before locating the bodies of the seven men who were "almost completely submerged."

They first located the bodies of Mine Captain Bill Tippett, Mine Inspector William Hill, track repairman Thomas Kirby Sr., trammer William Huot, and motor man Nels Hill. The remains of Trammers Arvi Wepsala and Henry Haapala were another 50-100 feet further away. Two of the men were pinned in the debris. Most were unrecognizable.

It was after dark when the remains of the men were taken to the surface. There, CCI Welfare Department Director William Moulton received the bodies and oversaw the process of laying out the men on the floor of the Morris dry, and gently rinsing them with a garden hose.

In his inquest testimony, Conibear said he had made rounds at Barnes-Hecker many times over the years, and knew many of the men by name.

When asked about the safety of the mine, Conibear said, "I have never heard a complaint from any source. As a matter of fact, on the day of that accident I would have been willing to go underground and made my bed there." On the day before the Barnes-Hecker tragedy, the mines were closed for election day, and Conibear had planned to go hunting with his friend, Captain Bill Tippett. Both men had suffered the losses of their fathers to mining. Tippett had been 13 when his dad succumbed from a mining-related lung disease in 1900. Conibear was 10 when his father died of injuries sustained in an accident at the Salisbury mine in 1887.

As boys, did both men have to go to work to help support their families? One can only wonder what common bonds were shared by the two friends, and what wrenching emotions washed over Conibear as he attended to the bodies of the dead.

Under Conibear's leadership, annual safety and rescue training took place at all CCI mines. In this photo dated 1916 at North Lake/Morris-Lloyd Mine, the ones who are identified are Tony Bucco at far left; front row right to left are Crooky (?) Anderson third from right, Bill Pascoe fifth from right, Jacob Korpi sixth from right, Art Carlson seventh from right; second row, Andrew Lahti seventh from right; third row, J. Henry Williams far right, Bill Tippett tenth from right; fourth row, Walter Tippett in gas mask second from right, Billy Williams sixth from right, Bill St. Onge seventh from right, Gust Erickson tenth from right; last row, Einar Johnson second from right, Poirier fourth from right, Jack (John) Wallberg fifth from right, St. Andre sixth from right. (Courtesy Tippett descendants.)

William Conibear was born on January 25, 1877, in Wales, England. He was three years old when his parents, William and Ellen (Butterfield) Conibear, immigrated to the United States. When the elder Conibear died in 1887, Ellen had the responsibility of raising six children ranging in age from 12 to one.

In 1911, when CCI initiated its Safety Department, Conibear was named the company's first safety inspector. In order to prepare him for this role, CCI sent him to Harvard for two years, and he received a degree in metallurgical engineering in 1913.

Among Conibear's responsibilities were data collection and analysis, safety training, first-aid training, and inspection rounds. In addition to monthly safety rounds by Conibear with the district superintendents, mine captains and the county mine inspectors, there were yearly rounds by teams of foremen and workers. Between 1913 and 1925, 99 foremen and 543 workers participated in these teams. Many improvements were implemented based on the teams' reports. The safety records provide interesting details for readers interested in learning more. Men who completed first-aid training and passed hands-on and written tests were awarded safety fobs with the CCI logo on the front and text on the back. In 1925, Conibear was Chairman of the Lake Superior Chapter of the National Safety Council. At the Lake Superior Safety Conference in Hibbing, MN, in August of 1925, he presented a paper entitled, "Metal Mine Accidents of all mines in the Lake Superior District" [sic]. At the National Safety Conference in Cleveland in September, he presented a paper entitled, "Safety Practices of the Cleveland-Cliffs Iron Company."

By 1926, Conibear had been employed at CCI for 33 years. He retired in 1942 and died in 1956. William Conibear and his wife Elsie were both born in England and married in Ishpeming. She predeceased him in 1943. They had one son, Dr. Chadwick Conibear of Benton Harbor, Michigan, and two grandchildren.

No one knows the private thoughts, emotions and sleepless nights that surely plagued the man in charge of the safety of the men employed by CCI. No one knows what anguished conversations he must have shared with his wife over the loss of friends and workers he had known for more than a decade. Conibear and several of the lost men were members of the Knights of Pythias.

Men who passed both hands-on and written first aid tests received a CCI safety fob. This one belonged to Walter Tippett. (Courtesy of Tippett descendants.)

Surely there must have been quiet personal gatherings of the CCI officials in the wake of the disaster, simply as men and friends, on whose watch catastrophe had happened in a mine regarded as one of the safest on the Marquette Iron Range.

(See also: Chapter 20: Catalyst for change, and Chapter 14: Lanterns in the dark)

YLEISET SÄÄNNÖT

CLEVELAND-CLIFFS IRON KOMPPANIAN

KAIVOS OSASTOLLE

Vahingon Loukkauksien
Välttämiseksi

Varovaisuus Tuottaa Varmuuden.

TOUKOKUULLA 1918

FINNISH

CCI published safety booklets in several languages including Finnish, Italian, and Norwegian, to name a few. (Courtesy of Tippett descendants)

16

Pocket Watch
Thomas Kirby Sr. (Trammer, 59) and Annie Kirby
Thomas Kirby Jr. (Skip tender, 23) and
Bertha Seablom Kirby

The Kirby family. Front, Jack, Edward, Richard; middle Annie Eddy Kirby, Thomas Sr.; back, Thomas Jr., Rita Kirby (Mrs. Charles Cowling), Joseph. At the time of the Barnes-Hecker tragedy, Edward was 12, and John (Jack) was a star football player for Ishpeming High School. (Courtesy Georgann Kippola Coon.)

When the Barnes-Hecker disaster was unfolding, Thomas Kirby Sr. was one of seven men on the third level who reached the underground passageway that connected with the Morris Mine. All were overtaken by a maelstrom of water and debris bursting through the underground workings of the mine.

At the same time, Thomas Kirby Jr. was one of four men climbing for their lives up the shaft. Neither father nor son would make it out alive.

The body of the senior Kirby was recovered on the afternoon of November 3, but his funeral was delayed while the family hoped for the recovery of Tom Kirby Jr. from the shaft. The younger Kirby was recovered on November 9.

Their funeral was held on Saturday, November 13, 1926. Because there had been a recent case of scarlet fever in the family, the funeral was held at the Kirby home instead of the church. The Reverend Lewis Keast of the Methodist Episcopal Church was the officiant.

The *Daily Mining Journal* reported that cars jammed Third Street in front of the house, and more than 1,000 people were at the Ishpeming Cemetery for the burial rites performed by the Sons of St. George.

Thomas Kirby Sr. was a talented musician who played the bass drum in local bands. The Ishpeming *Iron Ore* stated, "local bands will not be the same without this familiar figure beating the large drum."

He had immigrated from England in 1886, and became a boarder in the home of Martha Eddy and her children, one of whom was a young woman named Annie. Thomas Kirby (Sr.) and Annie Eddy fell in love and were married in 1896. Following the Barnes-Hecker tragedy, the Kirbys had four surviving sons: Joseph; John, who

Tom Kirby Sr. (Courtesy of Michigan Iron Industry Museum)

This photo ca. 1950 is believed to have been taken in the home of Rita (Kirby) and Charles Cowling. Front, twins Doreen and DeeAnn Kirby, Georgann Kippola (Coon), Beth Kirby. Back, John ("Uncle Jack") Kirby, Richard (Dick) Kirby, George Kippola, Edward Kirby. (Courtesy Georgann Kippola Coon)

was a star player on the 1926 Ishpeming football team; Richard, and Edward; a daughter, Rita (Ritta) Kirby Cowling, and two grandchildren. Kirby, who was a track cleaner at Barnes-Hecker, had previously worked at the Holmes and Lake Mines in Ishpeming and the Breitung in Negaunee. He was a member of the Sons of St. George.

Thomas Kirby Jr., who was 24, had worked at Barnes-Hecker for about a year as a skip tender in charge of the bell signals and loading chutes on the first level. Like his father, he was a member of the Sons of St. George. Thomas Kirby Jr. was a newlywed, having married Bertha Seablom on September 11, 1926. A son was born to Thomas and Bertha in 1927, and the baby was named Thomas after the father he would never know. Bertha became employed at the Gossard factory, but the time frame of her employment is unknown.

Kirby daughter Rita Kirby Cowling and her husband Charles felt the impact of the Barnes-Hecker tragedy on both sides of their family. Charles had lost his sister Lena Cowling Lampshire in 1925, leaving her husband, Richard Lampshire, a widower with four children to raise. When Richard Lampshire perished in the cave-in, Lena's parents, William and Esther Cowling, became guardians of Lena's and Richard's children.

In the years following the tragedy, several Kirby and Cowling descendants lived in close proximity to one another on East Empire, Vine, Maurice, and Michigan Streets in Ishpeming. The children were playmates, and the closely-knit families were able to support one another and celebrate family milestones and holidays together.

Doreen Britton, grand-daughter of Thomas Kirby Sr., said her father, Richard Kirby, was 15 when he lost his dad and older brother. She provided personal insight into the unspoken feelings of her father. She said the family had attended the 1972 dedication of the Barnes-Hecker monument on US 41. After the ceremony, they visited the mine site. "It was hard for him to be there," she said. When Doreen was growing up, the family would watch movies on TV, and one was about a mine cave-in with water coming in. Her father got up and abruptly left the room. "If only we could have had just one person talk about the tragedy, it could have opened some doors, and we could have had more information," Doreen said.

At the time of the Barnes-Hecker tragedy, two pocket watches and Captain Bill Tippett's notebook were recovered from the connecting drift with the Morris Mine. One of the watches belonged to Tom Kirby Sr., and in 2008, the watch was gifted to the Michigan Iron Industry Museum by granddaughters Doreen Britton and DeeAnn Truscott. Doreen said it was a logical decision to gift their grandfather's pocket watch to the museum, where more people would have an opportunity to see it.

The pocket watch of Thomas Kirby Sr. was found in the connecting drift between the Barnes-Hecker and Morris Mines

Time stopped for Thomas Kirby Sr. and Thomas Kirby Jr. in 1926, but their descendants are living timepieces who have kept the Kirby legacies alive for more than nine decades.

– Georgann Kippola Coon & Doreen Britton

(See also: Chapter 36: "The other side")

17

The youngest man
Arvi Wepsala (Vepsola) (18)
Trammer

Arvi Wepsala was only 18 years old, and the youngest man lost in the tragedy. A resident of Diorite, he was the sole support of his parents and siblings. Arvi was one of the seven men who were recovered from the drift (tunnel) between the Barnes-Hecker and Morris-Lloyd Mines in the afternoon of November 3.

In the early 1970s, Arvi's brother Sulo Wepsala had a contract with Cleveland Cliffs Iron Company to install fences and concrete

caps at abandoned mine shafts including Barnes-Hecker. Sulo hired his teenaged grandson Glenn Wing as a summer helper. Wing later said that in all the time he worked shoulder to shoulder with his grandfather installing the concrete cap on the Barnes-Hecker, not a word was spoken that Sulo's own brother Arvi Wepsala had perished in the tragedy.

– Glenn Wing

Photo courtesy Glenn Wing.

18

Connections
William (39) and Clara Huot Trammer

Huot brothers Oliver, Henry, and Peter in back and William in front. Their eldest brother Napoleon was not in the picture. (Courtesy Huot descendants)

The poignant story of William Huot's bereaved mother, Cecelia Huot, was told in the Marquette *Daily Mining Journal* the day after her son's funeral. Her family had delayed telling her the news of William's death in the Barnes-Hecker, because blindness and diabetes had left her in fragile health. They feared that news of her son's tragic death would be too much for her.

They asked her priest, Rev. Paul LeGolvin, to break the devastating news. He fought back tears as he struggled for the right words. She replied, "I know he has gone to provide a place with God for me," and slipped into a coma a short time later.

William was 39 and had worked at Barnes-Hecker for six years. His job as a trammer was to push heavy carts filled with ore. He was well known as a justice of the peace in Ishpeming Township. His body was one of the seven recovered from the tunnel connecting the Barnes-Hecker and Morris Mines.

William and Clara Huot on their wedding day, August 27, 1912. Their attendants were Peter Huot and his soon-to-be-bride Emma.
(Courtesy Huot descendants)

William and Clara Huot had seven children, the youngest of whom was only 10 months old. In the years after the accident, Clara used some of the compensation money from CCI to purchase a home. The 1930 census shows her as head of household on East Superior Street. The eldest daughter, Aurelia, 16, is listed as a seamer in a corset factory, which would have been the H.W. Gossard Factory in Ishpeming.

Clara Huot was one of the widows who didn't talk about the Barnes-Hecker tragedy, and in the nine decades since William's loss, her descendants have sought each other out. Douglas Anderson said the cousins have had large family reunions every other year since the 1998 publication of the book, *No* Tears *in Heaven; The 1926 Barnes-Hecker Tragedy*, by Thomas Friggens.

While the Huot descendants were trying to locate a missing cousin, the cousin was also wishing she could find the rest of her family. The only Barnes-Hecker history she had to pass along to her children were the book, *No Tears in Heaven* and the video, *Memories of a Misfortune*. As a follow-up to the 2016 Barnes-Hecker Remembrance, these members of the Huot family are now connected with each other.

There was another person in Belgium who was searching for information about the Huot family. As a member of the "Adopt a U.S. Tommy" program, Ronnyy Decuyper was assigned to tend the grave of Henry Huot, who died in Belgium in World War I. Henry was William Huot's brother. William's descendants were aware that someone in Belgium was caring for Henry's grave, but could not find out who it was. Through some connections, including the City of Ishpeming, Cliffs Shaft Mine Museum, the Barnes-Hecker Remembrance, and the Marquette Regional History Center, the gentleman in Belgium has been connected with the Huot family.

One member of the Huot family exclaimed, "This is unbelievable!" And indeed, sometimes the quest for connection leads us to people and places that are far beyond our wildest imagination.

– Douglas Anderson

A grace note was added to the Huot family story in 2023 when two of Henry Huot's great-nieces Cheryl Marietti and Susan Bitner traveled to Flanders Fields Cemetery in Belgium, where the keeper of Henry's grave, Ronny Decuyper, accompanied them to pay their respects at the grave of Henry Huot.

Photo courtesy of Huot descendants

19 The glue that held the puzzle pieces together
Nels (27) and Mae Hill
Motorman

Mae Hill found herself in the backseat of a car with her two small sons. Ahead of them was the hearse bearing the casket of her husband Nels. It was Saturday, November 6, 1926.

There had been a triple funeral at the Finnish Lutheran Church for Nels Hill, John Arvi Wepsala and Henry Haapala, all of whom had been recovered from the connecting tunnel between the Barnes-Hecker and Morris-Lloyd Mines. They were laid to rest side by side in the Ishpeming cemetery.

The Daily Mining Journal reported that hundreds of grief-stricken people were gathered outside of the church during the funeral.

Nels and Mae Hill
(courtesy of Anna Hill Mattson)

After the service, there was a long procession of vehicles going to the cemetery. Somewhere along the way, a string of cars took a wrong turn, and the hearse bearing Nels Hill's body followed. The car carrying Mae and her children followed the hearse, but the cars behind Mae remained with the rest of the procession.

A reporter misunderstood what he saw, and wrote, "One of the most heart-rending sights was the lone… shabby donated… automobile behind the hearse carrying the body of Nels Hill… The widow was without a friend or relative to console her in her sad plight."

One can only imagine how Mae and her loved ones reacted to that story. The reporter probably never forgot it.

On November 10, a correction ran with the headline, "Many paid tribute to the late Nels Hill." The article stated that there were actually 60 cars carrying mourners for Nels Hill in the funeral cortege to the cemetery.

Nels and Mae Hill had lived in the Barnes-Hecker Location

Ralph and Nels Allivian Hill at B-H Location summer 1926 (courtesy of Anna Hill Mattson)

across the street from shift boss Sam Phillippi. They had two young sons, Nels (4) and Ralph (2).

Nels was the motorman on the third level of the Barnes-Hecker Mine, where he had worked for two years. He was one of the seven men whose bodies were recovered from the tunnel that connected with the Morris-Lloyd Mine.

According to the 1920 census, Mae Hill came from a large family in Diorite. Her father Florien LaParche and brother, Edlore LaParche, were both listed in the Barnes-Hecker payroll, but neither worked the day shift on the day of the disaster.

In 1928 in Diorite, Mae married Wilfred LaForge, and in that same year she requested a lump sum from the CCI compensation funds to purchase a home.

Mae Hill LaForge with tenant Louis LaFreniere (courtesy of Anna Hill Mattson)

Mae and Wilfred later moved to the Witbeck Location near Republic, where they owned a 40-acre farm. They had three more sons: Lawrence, Paul, and Robert LaForge.

Mae's granddaughter Anna Mattson has fond memories of her little apron-wearing grandma who was never confined to the kitchen. Mae helped her grandkids find worms, took them fishing and berry picking, made smudge pots for warding off mosquitoes, and helped them roast marsh-mallows. During summer picnics at Sawyer Lake, Mae was the designated pitcher for ball games. When her team wasn't at bat, she played bingo. Anna said Mae had a large garden and a cold cellar for storing canned goods.

Anna and her cousins would be dispatched to the neighboring Janofski farm to pick up bottles of milk, and Mae taught them not to shake the bottles because the cream would float on top of the milk, and she used the cream for shortcake. Mae earned extra money by allowing neighbors to harvest hay and timber on her land.

Anna's dad Nels A. Hill and his uncles took Christmas trees to the Detroit area where relatives had a tree lot. Those relatives would come to stay with Mae during hunting season, and to visit during the summer.

"Her door was never locked, her refrigerator never empty, and she always welcomed everyone who walked through the door. Family visits were her happiest times," Anna said. "She was the glue that held the puzzle pieces together."

At an event at the Michigan Iron Industry Museum in the 1980s, Anna learned that one of her dad's uncles, Walter Hill, was still living. Anna's family helped the two elderly men get together a couple of times, and both took great pleasure in each other's company.

"In spite of the losses and incomprehension of what those days around the Barnes-Hecker tragedy must have been like, I think we have all come away with a sense of gratitude, empathy, and love for the ones we never knew," Anna said.

– Anna Hill Mattson

20 Catalyst for Change
William F. (43) and Nellie Tippett
Mine captain

William F. Tippett became captain of the Barnes-Hecker in 1917, when shaft-sinking was in progress and buildings were under construction. In the fall of 1918, he and his wife, Nellie, and their children, William G. and Margaret, moved into the newly completed captain's house in the Barnes-Hecker Location that consisted of ten duplexes and the captain's house.

Bill married Nellie Trevarthan in Quinnesec in 1906. Both of their children were born in Calumet, Michigan, where there were thriving copper mines. Bill and Nellie experienced the anguish of parents whose son was born prematurely, and his survival tenuous. Baby Will was so

Captain Bill Tippett

tiny and fragile that his mother wrapped him snugly in a shoe box and placed him near the open oven door to keep him warm. By the time the boy was five, the lad had endured polio and several surgeries on his legs.

North Lake Band, 1914: Front row left with tuba is Bill Tippett, Jim (last name illegible), Chet Uren, Bill Tippett Jr.; row 2 unknown, unknown, Walter Tippett with cornet, Emil Erickson, Jon Truscott, Alf Tamblyn, Joe Labouef, Harry Sayle; row three Albert Ericson, Emil Ericson, Jack (John) Wills, Jack (John) Wallberg, Harry Scarffe, unknown. (Courtesy of Jack Deo)

Will is pictured in a 1914 photo of the North Lake Band, where his dad played the tuba.

In 1915-16, when Bill was a shift boss at Morris-Lloyd (North Lake) Mine, he completed an 18-month curriculum offered by CCI to train men for advancement in underground mines. When he took the helm of Barnes-Hecker, he became known as one of the most respected mine captains in the district. In the wake of the tragedy, the Ishpeming *Iron Ore* stated, "He was familiar with every foot of the workings and was considered one of the most skilled men in the company's employ."

When catastrophe crashed into the Barnes-Hecker on November 3, 1926, Bill Tippett died with 49 of his men and the county mine inspector. The tragedy also claimed his brother Walter Tippett, and narrowly missed killing brother Albert and half-brother Rutherford Wills. Bill's body was recovered from the connecting drift (passageway) between the Barnes-Hecker and Morris-Lloyd on the afternoon of the cave-in. Bill's brother Tom was captain of the Lloyd at the time.

Bill and Nellie Tippett in Barnes-Hecker Location, spring 1926. (Courtesy Tippett descendants)

The magnitude of grief that swept up the Tippett families was immeasurable. Nellie Tippett was left to console her daughter, 16-year-old Margaret, who lost both her dad and her fiancé Joe Mankee.

When the mine collapse occurred, the boy who had had polio was 18. Will Tippett had chosen to become a bookkeeper, and helped to support his mother and Margaret after the tragedy. Nellie and Margaret moved to northern Wisconsin, where Nellie had relatives. Margaret later married and had a large family. Will moved to Flint, where the booming auto industry held countless opportunities. He became employed by several different companies that supplied the auto industry. He and his wife Lila had two children.

In the latter years of Will's life, post-polio syndrome robbed him of the use of his legs. Then, grandson Eric William Tippett ("Will") moved in with his grandparents, and lifted his grandfather into and out of bed each day for the last four years of his grandfather's life.

"Will" retired from a career in the Army in 2018 and now works in private industry. Two of his most precious possessions are the straight razor and shotgun that were used for the last time in the hands of his great grandfather, mine captain Bill Tippett. He remembers his great grandmother Nellie's fragrance of ylang-ylang, and her quiet and kind manner.

Lila and William G. Tippett, Flint MI ca. 1940s (Courtesy Eric William Tippett)

Will remembers his great aunt Margaret as a fun and charming woman who was sharp-witted and wickedly funny. He said the extended family often talked about the Barnes-Hecker tragedy with grace and forbearance.

Will Tippett has pondered the tragedy and its aftermath for many years. He describes it as a catalyst that shocked people into thinking differently about their lives. He said the accident pinpoints a pivotal time in history that marks the great migration from rural life to urban industrial places like Detroit and Flint. "It looks like a history of America," he said.

–Eric William Tippett

Bill Tippett's pocket watch is in the collection of the Michigan Iron Industry Museum. As with many aspects of the Barnes-Hecker, there are two different narratives about this watch. The museum caption states the watch was retrieved from Tippett's pocket, and this information would have come directly from Tippett's family. The Mining Journal reported that the watches of Tippett and Kirby were found with Tippett's notebook in the connecting drift between the Barnes-Hecker and Morris Mines a few days after the bodies were recovered. (Photo courtesy of Michigan Iron Industry Museum)

See also: Chapter 26: A grateful heart; Chapter 12: Lone survivor; and Chapter 41: Legacy bearers

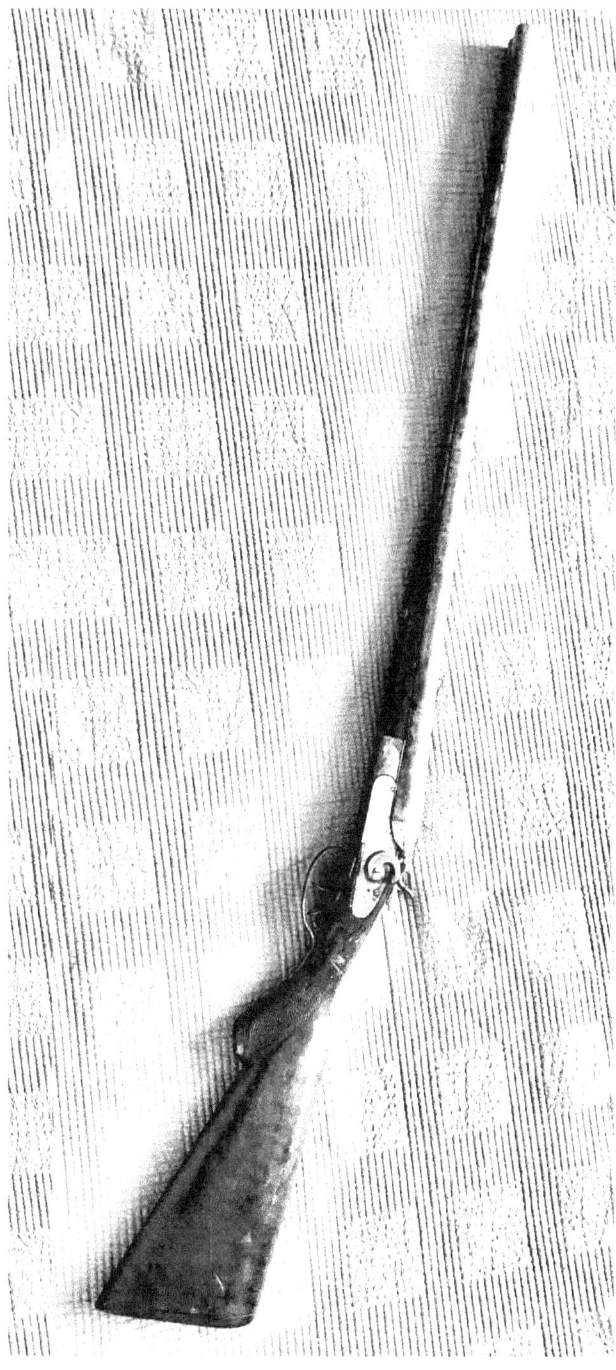

Bill Tippett's shotgun was fired for the last time in Bill's own hands in 1926. (Courtesy Eric William Tippett)

Cardboard Boxes
William (39) and Elizabeth Hill
Marquette County Mine Inspector

On Election Day, November 2, 1926, William E. Hill won re-election for a second two-year term as Marquette County Mine Inspector. He was well respected and well acquainted with mining operations across the spectrum on the Marquette Iron Range.

He had worked underground at Salisbury Mine, and participated in an 18-month curriculum of classes offered by CCI in 1914 and 1915, for men interested in moving up into supervisory roles. Hill finished the classes with high grades, and then left CCI to go into business.

Eventually, he ran for public office and was elected to his first term as mine inspector in 1924.

(Courtesy of Patricia Hill Makinen)

Elizabeth Ahola Hill and Ida Maki (courtesy of Don Hill)

Hill had planned to celebrate his re-election victory by going rabbit hunting on Wednesday, November 3. His son Norman remembered seeing his dad's gun leaning in a corner by the telephone. When the phone rang early that morning, Hill readily agreed to change his plans and fill in for North Lake District Superintendent Charles Stakel on routine underground rounds at the Barnes-Hecker Mine.

By noon, Elizabeth Hill learned that there had been a cave-in at the mine, and her husband was among the men trapped underground

William Maki and William Elias Hill (courtesy of Don Hill)

William Hill family: Ellen Hill Luoma Kosten; Elizabeth Ahola Hill; Norman Arthur Hill (b. Aug. 28, 1914); Emil Adolph Hill (b. Nov. 5, 1915.); William Arnold Hill (b Aug. 4, 1913).
Courtesy of Don Hill

and presumed dead. Her husband's body was one of ten recovered from the mine.

Because Hill was an elected official rather than an employee of Marquette County, he was not eligible for workmen's compensation benefits; however, at its November meeting, the Marquette County Board stepped in and voted unanimously to pay Hill's November and December salaries to his widow Elizabeth.

In addition, CCI President William G. Mather felt it was essential to provide Elizabeth Hill the same compensation as the other widows of the lost men, totaling $8400, or $28 per week for five years.

Elizabeth and William Hill had four children: Arnold, 14; Norman, 12; Emil, 11; and Ellen, 9. After the compensation expired in 1932, Elizabeth was supported by her children, and later lived with her eldest son Arnold in California and daughter Ellen in Florida. In the 1950s, Elizabeth traveled by train to visit her sons Emil in Ishpeming and Norman in Sault Ste. Marie, in Michigan's Upper Peninsula.

Norman's daughter Patricia Hill Makinen has fond memories of her grandmother's two-week visit. She was 12 at the time, and recalls that her grandmother arrived with simple parcels of luggage consisting of two cardboard boxes tied with twine. She described Elizabeth as a pleasant, petite woman with intense blue eyes and her hair pulled back in a bun. Elizabeth taught Patricia to knit, crochet and make rag rugs. Patricia said her grandmother had a survivor's fortitude and liked to feel useful. One night, she heard Elizabeth get out of bed and begin mopping the kitchen floor. Elizabeth Hill didn't talk about the Barnes-Hecker tragedy, and Patricia feels that her dad, Norman, lived his life with unresolved grief. Norman once told his son Jim not to get too close to anyone, because they might leave.

Elizabeth left important legacies to her descendants that are encompassed by the Finnish word "sisu," which translates into a combination of resilience, fortitude, grit, stoicism, courage and grace under pressure.

During the summers, when Patricia travels from California to Michigan's Upper Peninsula, she visits her grandfather's final resting place in the Ishpeming cemetery. By playing music from Sibelius's *Finlandia* on her phone, she honors the Finnish heritage of her grandparents, and the quiet, cheery strength of her tiny little grand-mother.

– Patricia Hill Makinen

See also: Chapter 15: I would have made my bed there; Chapter 20: Catalyst for change

Casket of William E. Hill leaving the Finnish Lutheran Church. Photo from *Chicago Evening American* (Courtesy of Don Hill)

22 A son was born
Jack (23) and Florence Hanna
Brakeman

Jack Hanna was one of the three men who were swept away as they fled up the ladder behind Rutherford Wills.

Jack and his wife Florence (Verran) were born in Ishpeming and married in Detroit on November 17, 1925. They returned to Ishpeming in July of 1926, and he started working at Barnes-Hecker three weeks before the tragedy.

Jack was also survived by his parents, Mr. and Mrs. Thomas Hanna of Molalla, Oregon, a sister Mrs. Howard Billings of Petoskey, and a brother William Hanna of Detroit. Jack had an aunt and uncle, Mr. and Mrs. Thomas Clague, in the Salisbury Location (neighborhood south of Ishpeming).

The funeral was held on Thursday, November 11, in the home of Captain John T. Verran, on the corner of Third and Empire Streets, with the Reverend Lewis Keast of the Methodist Episcopal Church officiating.

John J. Hanna from *Mining Journal* (Nov. 12, 1926)

Florence Verran Hanna was imminently due to deliver the couple's first child. Baby Jack was born twelve days after the Barnes-Hecker tragedy, on Monday, November 15, 1926.

Florence and young Jackie remained in the Verran home, and she was one of the Barnes-Hecker widows who eventually worked at the Gossard women's undergarment factory. She suffered another unspeakable loss in 1939, when twelve-year-old "Jackie" was riding his bike, hit a car, and died from his injuries. Jackie was laid to rest close to his father.

Although there are no direct descendants left to pay tribute to Jack, Florence and Jackie Hanna, they will always be remembered among the people whose lives were caught in the maelstrom of the Barnes-Hecker tragedy.

23 No Warning
S.R. Elliott (1874-1946)
Mining Manager, Cleveland Cliffs Iron Company

S.R. Elliott was in Crystal Falls when he received word that a serious accident had occurred at Barnes-Hecker. Elliott was the assistant general manager at CCI, and he had laid out the plans for the mine. When he arrived at the mine around 2 o'clock, he learned that water and mud were 185 feet from the surface, and he knew immediately that no one underground could be alive.

Elliott's sworn testimony from the coroner's inquest provides salient answers to questions that have lingered for more than 90 years. He said that before work began on the Barnes-Hecker, there had been consultations with the superintendents of other mines on the Marquette Iron Range about the safest way to approach the construction of the mine.

In his inquest testimony, Elliott said the ground in the area of the Barnes-Hecker was like a sponge. That fact was known from coring samples taken while the territory was being explored, long before shaft-sinking began in the fall of 1917. In 1918, the Ishpeming *Iron Ore* reported, "there was 184 feet of quicksand in the vicinity of the ore body." The paper reported that in such a situation, it would have been possible to freeze the ground in order to sink the shaft, but at a price tag of

S. R. Elliott (Courtesy of Michigan Iron Industry Museum)

a million dollars, the shaft was sited 2,000 feet from the ore body instead.

Elliott had been the superintendent of the Negaunee District when the shafts were sunk for the Negaunee and Maas Mines in the early 1900s. Both of those mines were wet, soft ore mines with similar geology to the Barnes-Hecker. The similarities among these mines, along with the Stephenson Mine in Gwinn, were noted by State Geologist L.P. Barrett in his inquest testimony.

As to the question of whether or not there was a swamp above the workings, Elliott said there was not, and even if there had been, it would have been irrelevant because numerous mines were operated beneath swampland. In fact, when asked about how wet it was in the uppermost workings at the time of the cave-in, he said, "it was dry as a bone. We could not have operated our *scrapers* in wet territory because wet ore would have clogged the chutes."

When the cave-in occurred, it left a funnel-shaped crater 200 feet by 300 feet wide and 60 feet deep, with an eight-foot opening at the bottom. Entire trees had been swallowed up, and Elliott estimated the volume of lost ground at 200,000 cubic yards. Elliott said the north side of the crater was sandy, high and dry and the south end was muskeg

The cave-in began to fill with water, and pumps were put into operation on the afternoon of the tragedy. For every foot that the water was lowered, it rose two feet. Nevertheless, there was hope of draining both the crater and the mine shaft.

Elliott said that after the tragedy, there was a slim chance that the pumpman might have survived in the water-tight pumphouse, but that hope was dashed because the air line in the shaft was found to be severed, and the concrete pump house could not have withstood the weight of an entire shaft full of water.

Elliott's testimony about pressure is cited verbatim to assure that the language is technically correct:

> There was no possible chance for anybody to be alive in the mine. They could not live because the pressure would be so great, even if they were held in an end of a drift where the air

was compressed, due to the height the pressure of the air would
be so great with practically a 600-foot head to have caused

This photo shows what the caved ground looked like on the afternoon of the tragedy, November 3, 1926. In the background and on the left side are snow-capped layers of ground that fell away when the ground failed. In the foreground are old tree stumps that became submerged after water filled the crater. A description that ran in the Friday, November 5, Mining Journal said, "A peculiar situation has arisen yesterday [November 4], when it was discovered that the cave-in, which was free of water on Wednesday, had filled to within 20 feet of the surface."

almost instant death, and I knew by rough calculations which were made very early that afternoon that practically every space in the mine was filled.

As pumping continued in the shaft, Elliott said they encountered "the shaft absolutely filled with sand and boulders and refuse of all kinds." Elliott also said the exposed sand in the shaft was hard packed. Round-the-clock pumping resulted in temporary gains before water would once again fill the shaft. Pumping at the cave-in also resulted only in temporary progress. On November 12, *The Daily Mining Journal* reported, "A piece of a root, about four inches in diameter, was brought to the surface Friday by the bailer, indicating, officials said, that surface material has been carried from the caved crater to the shaft." [Author's note: Might the tree root have been four feet in *length* instead of *diameter?*]

As the days passed, Elliott had growing concern for the safety of the men working to repair the shaft. In his inquest testimony he said, "There was no one that could tell at what instant the water and sand… in the mine would seek its natural level and rise up 575 feet in the shaft. That would mean that every man who was working in that shaft would be instantly killed."

From the day of the cave-in, Elliott had been in continuous communication with the leading hydrology and mine stripping companies in the country. After weeks of pumping with minimal progress, he telegraphed experts and asked them to come and assess the situation.

On November 20, the western manager of the Foundation Company, Mr. Adgate, arrived from New York. Elliot said the Foundation Company was the foremost hydrology company in the country and perhaps the world. Adgate visited both the mine site and cave-in and studied the maps.

"I stated the conditions to him and told him this thing had worried me so much," Elliott said.

Adgate replied, "You know what I would do? Do you think I would allow a man to go down into that shaft? I wouldn't think of it for one minute."

"I turned around and rang the telephone, called the mine and since November 20, there has nobody been in that shaft," Elliott said. *The*

This photo shows that water ultimately filled the cave-in to within a few feet of the high sandy bank on the north shore. The man in the photo serves as a reference for the size of the lake. To this day, the north bank is hard-packed sand that rises about four feet above the clear water. Although there were claims that a swamp had caved in, and many headlines repeated the inaccuracy, the photos and other documentation belie that claim. (Both photos courtesy of Marquette Regional History Center)

Daily Mining Journal reported that within 20 minutes after the men were ordered out, the water rose 80 feet in the shaft.

Pumping continued until January 11, 1927, when the effort was deemed futile and CCI had to declare the mine abandoned.

Elliot said that when they began opening the main levels of the mine, they encountered a great deal of water on all three levels. In 1922, North Lake was drained and a trench was dug to reroute surface water away from the mine. As a result of the efforts to de-water the mine, pumping volume decreased from 3000 gallons per minute to 700 gallons per minute, a volume that was handled easily by the main pump. According to the CCI annual reports, the pumping volume remained fairly constant thereafter.

"We knew we were working under heavy ground," Elliott said. "We knew there was water, but we were not afraid of it. We had built our dams and were ready to close them on a moment's notice. But there was *no warning*." [Emphasis added]

Following the cave-in, Elliott said there was frustration over the inability to identify where the water could have come from. He was adamant that the catastrophe at Barnes-Hecker was peculiar to that mine, "Just some absolutely unforeseen condition which has made this terrible disaster. It is no ordinary condition; it is something that we knew nothing about."

Stuart Rhett Elliott was born in Beaufort, SC on April 26, 1874. He earned his bachelor of science in metallurgy from Lehigh University in Lehigh, PA, in 1897. He became an assistant mining engineer for CCI in Ishpeming in 1898, and later completed his master of engineer of mines.

His employment record included serving as superintendent of the CCI-owned Crosby Mine in Nashawauk, MN, from 1902-1904, then as superintendent of mines for CCI in Negaunee, MI, from 1904-1906. He then worked for a few months as mining manager for Bethlehem Steel in Cuba before returning to CCI.

During World War I, he began service in the U.S. Army in August 1917. With the rank of major, he was sent overseas in charge of an engineering company. While in France, he became a lieutenant colonel. After his discharge from the Army in 1919, he returned to CCI. In January of 1927, he was named assistant general manager, and six months later became manager in charge of all CCI mining operations.

He married Bessie Reid of Baltimore MD on December 1, 1919, and she survived him.

A member of many mining professional organizations, Elliott was a director of the American Institute of Mining and Metallurgical Engineering, president of the Lake Superior Mining Institute, and board member of the Michigan College of Mining and Technology [now Michigan Tech University].

His civic involvement included serving as president of the Ishpeming hotel company that operated the Mather Inn; the Ishpeming board of public works; the Marquette County Road Commission; director [board member] of the First National Bank of Negaunee; and a senior warden of the Episcopal Church.

24   A Perilous Climb
W.W. Graff (1877-1944)
Gwinn Mine Superintendent

Since the day of the accident on November 3, efforts to pump water out of the shaft had progressed well, dropping the water level almost to the top of the first level, which was 600 feet from the surface. A hard-packed mass of sand, timbers and debris was jammed in the shaft 420 feet from the surface.

On November 7, four officials went down for a closer look. CCI Safety Director William Conibear, North Lake District Superintendent Charles Stakel, Gwinn District Superintendent W.W. Graff, and Maas Mine Captain Joseph Thomas of Negaunee went down to the top of the mass wedged in the shaft. There was enough room along the side of the mass for a man to make a treacherous 75-foot descent to assess the situation. Graff took on the challenge.

The Daily Mining Journal's description of Graff's climb was vivid:

> ...the additional 75 feet was negotiated by... Graff... who made a hazardous climb down the broken ladderway and squirmed through 20 feet of matted timber choked in the shaft. Graff declared that he could still see another 20 feet of timber below him, but could hear no water sounds. He estimated that water had dropped almost to the first level.
>
> The timber torn loose from the sides of the shaft and brought upward from the levels below is jammed tight, forming a wedge 40 feet thick and held so securely that the force of gravity which would otherwise drop it down the shaft has no effect on its ponderous weight
>
> The shaft is lined with concrete and the butt ends of huge timbers were imbedded [sic] in the sides. These timbers were

broken off like toothpicks by the surging inrush of water and sand and the task of setting new ones is extremely difficult.

After November 7, water dropped and rose in the shaft, and crews of men were able to remove the mass of hard-packed sand, timber and debris almost to the first level, the floor of which was 600 feet from the surface. A two-and-a-half ton boulder was also blasted out of the shaft.

For a time there was hope that pumping efforts could be successful and the first level could be reached. But as the month of November progressed and water levels continued to rise and fall in both the shaft

"Relief work futile – photo shows miners repairing shaft, following accident at Barnes-Hecker Mine. There is little hope of finding bodies of dead men until flood waters recede." *Chicago Herald and Examiner*, undated. (Courtesy of Tippett descendants)

and the crater of the cave-in, CCI Assistant Superintendent S. R. Elliott had growing concern for the men's safety. On November 20, he ordered all men out of the shaft, and twenty minutes later, water rose 80 feet in the shaft. Pumping efforts after November 20 proved futile, and in January of 1927, the mine was declared abandoned.

Wilbur Wilson Graff was born in Lehigh, Pennsylvania on July 20, 1877. He graduated from Lehigh University and joined the CCI Engineering Department in November 1901. In just three years, he received a promotion to superintendent of the Cliffs Shaft and Moro Mines. In 1909, he was superintendent of the North Lake District Mines, and became superintendent of the Gwinn District in 1916.

During Graff's tenure in the North Lake District, the 1909 shaft-sinking for the North Lake Mine was underway by a contractor using the Caisson Method to sink the shaft through a swamp. In that same year, the Ishpeming *Iron Ore* printed the proceedings of the Lake Superior Mining Institute including a paper by Frederick Adgate of the Foundation Company, describing the Caisson Method of shaft sinking that was used at the Swanzy Mine in Gwinn. In 1926, Adgate became one of the hydrologists who came to assess the disastrous situation at Barnes-Hecker.

When Graff was superintendent of the North Lake District, there was a Barnes Mine that operated between 1910 and 1912, several hundred feet northwest of the Barnes-Hecker site. The only available records of the existence of the Barnes Mine are in the 1910 CCI annual report and the 1912 financial report.

In 1927, he became superintendent of the Negaunee and Athens Mines, and in 1940 became superintendent of the entire Negaunee District comprising the Athens, Negaunee, Maas, and Cambria-Jackson mines.

Graff was associated with the Marquette Range Engineers Club, Lake Superior Mining Institute, and the American Institute of Mining and Metallurgical Engineers. He was also an officer of the Episcopal Diocese of Northern Michigan.

Graff was 67 when he died in 1944.

See also: Chapter 51: The Barnes before the Barnes-Hecker

25 Coring samples and a cork
Leslie P. Barrett (1887-1972)
Mining Geologist and Appraiser of Mines
State of Michigan

State geologist L.P. Barrett was traveling when he received word of the cave-in at Barnes-Hecker. He arrived in Ishpeming a few days later.

Barrett had become a geologist for the Michigan Geological Survey after graduating from the University of Michigan in 1913, and in 1920, became the mining geologist and appraiser of mines for the State of Michigan.

Barrett was known for his fair approach in dealing with the mining companies, local communities, and the state when it came to tax valuations for each mine. The taxes to be levied were dependent on the geology and the estimates of marketable ore that the mines could expect to produce.

Those estimates, in turn, depended on coring samples, of which Barrett said he had thousands in his office. The coring samples were numbered and matched with detailed profile and overhead maps provided by the mining companies. Each coring sample revealed what kind of ground there was on the surface, what each layer of substrate was, how thick each layer was, and

L.P. Barrett from *The Daily Mining Journal* on 11/12/1926. (Courtesy of Tippett descendants)

the depth of ore deposits and their quality. The coring samples also identified the likely geographic area, from a bird's-eye-view, where the limits of the ore deposits were located.

In his testimony at the February 1927 coroner's inquest, Barrett said that geologists who had studied the topography of the area around the Barnes-Hecker believed that it was formed by pre-glacial drainage, with some of the ore deposits being in topography that was very irregular. There was discussion about the geology above the workings of the Barnes-Hecker, and Barrett said that based on coring samples, there was an average of 230 feet of ground above the workings of the mine, and at the time when mining began, the ground was solid.

In his inquest testimony, Barrett said there were known situations in U.P. mining districts where coring samples had missed voids and fissures that could only be found from underground. (Inquest, p.48) Such fissures were known to be repositories for water.

He said he was not offering the information as a cause of the Barnes-Hecker tragedy, but simply citing that such unusual situations were known to exist.

"I don't know what caused that accident," Barrett said. "It seems inconceivable to me that that this ground could have... caved through that mat." Barrett was referring to the "mat" of cris-crossed timbers and logs that were placed into mined-out spaces before they were blasted down flat, as described by Al Miners. Barrett's statement meant that he did not believe that the cave-in could have originated from above.

When questioned about whether North Lake could have contributed to the cave-in, Barrett stated flatly, "It couldn't have. It was empty."

Barrett was asked specifically about whether a swamp was present above the workings of Barnes-Hecker, and he replied, "No, anyone can see that because the place where it actually caved in directly over the ore body. . . is sand with hardwood timber on it."

Barrett said his knowledge of the geology of Barnes-Hecker and the challenges of mining it were discussed many times with CCI officials ever since the beginning of the mine. He had been on-site at the mine numerous times prior to1926, and was there in 1922 when a drainage ditch was dug to drain water away from the mine site. Over the years, he said he had also made numerous reviews of the maps.

When asked his opinion of the caving method of mining used at Barnes-Hecker, he replied that it was the only method that could be employed with safety. Barrett noted that the only other possibilities would have been either open pit or stoping, both of which he deemed too dangerous.

When asked his opinion about whether anything that could have been done from a safety standpoint was not done, Barrett replied, "No, I think rather unusual precautions—when I say unusual I mean in comparison with my experience in other mines—were taken here, particularly in the amount of timber [matting] that was put in this top sub[level]."

Barrett was asked his opinion about whether the Barnes-Hecker could be safely pumped out and/or accessed from the drift between it and the Morris-Lloyd. He replied that he concurred with S.R. Elliott's assessment that the head of pressure created by the water levels in both the mine and the cave-in couldn't be disturbed without creating risk of a sudden and uncontrollable incident that Barrett likened to pulling a cork and having the contents explode instantaneously. More lives would be lost in the attempt.

In a statement given to *The Daily Mining Journal* of November 8, 1926, Barrett said,

> I find that the precautions taken underground to mine out the ore body above, where the cave-in occurred, were adopted after a most careful discussion and study of the situation with the entire operating staff of the company and I am of the opinion that throughout the entire length of the Lake Superior District the mines are being operated under similar precautions.
>
> The precautions taken and mining methods used were such that there was no reason to suspect it possible for such a catastrophe to occur. In short, the officials of the Cleveland-Cliffs Iron Company took all possible precautions to insure [sic] the safety of their men and no blame can be attached to the company or to any individual for what occurred.

During Barrett's career, he was known for his interest in pre-Cambrian geology and his oversight of field studies for instructors and students of geology in Michigan, Wisconsin, and Canada. Those expeditions took place in the vicinity of mining districts and resulted in creation of maps and information used as reference material later on.

After Barrett died in 1972, the Geological Society of America published a memorial leaflet, in which author R.M. Crump said that Barrett's field expeditions were preceded by months of literature search as part of his painstaking preparations. Crump said,

> He was a natural geology teacher, one who believed one learned geology best and most readily in the field... His associates were impressed by his scientific approach and dedication to geological problems which were encountered. His enthusiasm and energy were infectious and we all worked a little harder because we had a good role model. It is the boast of many that they once worked for L.P. Barrett.

Leslie P. Barrett was born in Boston in 1887, and earned his bachelor of science degree from the University of Michigan in 1913. He began his career with the Michigan Geological Survey before serving from 1917–1919 with the First Division, American Expeditionary Forces during World War I in Europe, leaving military service with the rank of captain. He became state geologist and mining appraiser of iron and copper mines in 1920.

In his role as appraiser, his study of coring samples at the Tracy Mine in Negaunee indicated that a drill had wandered off course and missed a million tons of taxable ore. He was proven right when the adjacent ore body was mined and a lost core barrel was found. Because of that experience, he began working for the Tracy Mine parent company, Jones and Laughlin, in 1927. He retired as chief geologist—vice president of iron ore mining subdivisions in 1952. During his tenure at Jones and Laughlin, he found and developed the Benson Mines in Star Lake, NY.

Following retirement, he was a geological consultant to the Atomic Energy Commission, Ashland Mining Company, and McLouth Steel Corporation. Barrett's writings were published in several professional journals.

He was a member of the Geological Society of America, the American Association for the Advancement of Science, the Society of Economic Geologists, and the scientific society of Sigma Xi. He was also a member of the University Club of Chicago, Chippewa Club of Iron Mountain, and the New York Last Man's Club of the Fifth Artillery, First Division.

Barrett was a resident of Negaunee, Michigan, and survived by his wife Leona Locher, son Donald C. Barrett of Seattle, two brothers, and a grandchild.

See also Chapter 23: No warning

26   A Grateful Heart
Walter (32) and Marian Tippett
Miner/Stemmer

Walter Tippett had become a hero in 1924, when he was the night captain of the Marquette Police Department. An attempted armed robbery led to the deaths of Police Chief Martin Ford, his son Lloyd, and patrolman Thomas Thornton. Walter was the officer who shot and killed the assailant, Oscar Lampinen, as he attempted to swim across the Chocolay River. In the midst of the gunfight, a bullet narrowly missed Tippett, and his son Dewey recalled being a small boy as his dad lifted him up to put his finger into the bullet hole in the tree where Walter had stood.

Recovery of the body of assailant Oscar Lampinen in August of 1924. Walter Tippett, who was the night captain of the Marquette Police Department at the time, is pictured. (All photos in this chapter courtesy of Tippett descendants.)

Walter was hailed as one of the best wrestlers in Marquette County. His last match was on September 17, 1926 at Liberty Hall. The match had been rained out of the Marquette County Fair.

The fact that Walter had taken a man's life weighed heavily on him, and he left the police force to take a job at the Marquette Branch Prison. Along with harrowing incidents, Walter had opportunities to prevent volatile escalations and conflicts.

Dewey related an incident about a confrontation that occurred after a foiled riot. An inmate in the barbershop had vowed that he was out to get Tippett for his role in the incident. Walter coolly walked into the barber shop, waited for the inmate's barber chair to open, took a seat, and

Wedding photo of Walter and Marian Bengry Tippett.

asked for a shave. "When the inmate picked up the [straight] razor, Dad said, 'By the way, I hear you are out to get me? Now's your chance.' When Dad called his bluff, the guy began to tremble so bad—that was the end of the threat and the shave."

A well-known athlete, Walter "Tarzan" Tippett played center on the Ishpeming-Negaunee Allstars football team that played the Green Bay Packers in the Packers' first road game in October 1919. He was also touted as being one of the best, if not the best, wrestler in the U.P. His last wrestling match was in September of 1926

First and foremost, Walter was a family man. He and his wife Marian had known the devastating loss of their first two children, three-week-old Van who died in the 1918 flu pandemic, and three-year-old Vivienne who died of tubercular meningitis in June of 1919.

The joy of children returned when Dewey was born in 1921. They adopted Irene in 1922, and by 1926, Walter and Marian had four children: Irene (11), Dewey (5), and two-year-old twins Dawne and Dixie. The family had moved from Marquette to Ishpeming in the

Marian and Walter Tippett with daughter Irene (standing), Dawne, Dewey, and Dawne's twin Dixie, ca. 1924.

spring of 1926 to help Marian's parents, Ernest O. and Emma Bengry. Emma's health was declining, and she passed away in July of 1926.

The family remained in the home of Marian's father, who was a bookkeeper for the Lake Superior and Ishpeming Railroad. Marian's half-brother Ernest G. Bengry was 12, so Marian ran the household. As fall approached, Walter wanted to avoid the winter commute to the prison, and he plied his brothers, Mine Captain Bill Tippett at Barnes-Hecker and Mine Captain Tom Tippett at the nearby Lloyd, to hire him at one of the mines. Walter was no stranger to mining, having worked at the North Lake /Morris-Lloyd Mine, where he was one of the men who completed an 18-month series of supervisor training classes offered by CCI in 1914-15.

On the morning of November 3, 1926, he joined three of his brothers, Bill and Albert Tippett, and half-brother Rutherford Wills, for the *cage* ride underground at Barnes-Hecker. Step-dad Jack Wills was the "dry man" who managed the locker room building on surface. When disaster struck, Walter and Bill were caught underground. Albert had gone to the surface about 10 minutes before the tragedy, and Rutherford was the only man to escape the flooding mine. Bill's body

This photo was taken in the Barnes-Hecker Location in the spring of 1926. From left, Jack (John) Wills, Walter Tippett, Tom Drew, Capt. Bill Tippett and his son Will Tippett. The boy in front is Walter's son Dewey.

was retrieved on the afternoon of the tragedy, but Walter was never recovered.

In the wake of the tragedy, Marian and the children remained in the home of her father in West Ishpeming. She wrote a letter to William Moulton, who ran the welfare department at CCI, to express her gratitude to the company for the compensation provided to the widows of the lost men. In the letter, dated December 3, 1926, she wrote, in part:

> Walter absolutely knew no fear, and he has gone as he wished. On his legs, like a man; and I wish to be as brave as he was, and face whatever comes to me. But it would be much harder to stand up under, were it not for the help I am receiving from the CCI Co.

The days are getting harder to bear on account of the loneliness, and the nights are practically sleepless. When sleep does come, waking only brings stern realization and a renewed battle for strength to withstand the grief. I will try my utmost to live up to one of Walter's sayings, "I may be down, but I'm never out."

She signed the letter, "With many thanks from a grateful heart, I am sincerely, Mrs. Walter Tippett."

Marian's grateful heart remained with her for the rest of her life. In 1930, she married Russell Hill, and they celebrated their 25th anniversary before cancer claimed Russell's life. They knew the joy of having most of Marian's grandkids living on the same block in West Ishpeming, and Russell was the only grandfather the children ever knew.

Through the years, the Barnes-Hecker has been a frequent topic of conversation, and family members have visited the mine site and the lake that formed in the cave-in, where they tossed garden flowers onto the sparkling water in remembrance. Marian and Dewey are known to have attended the 1972 dedication of the monument commissioned by Ely Township, and numerous family members in three generations attended the monument rededication at the Michigan Iron Industry Museum in 1988. Dawne made remarks at the dedications of monuments in 1988 and 2002. A family member attended the 2011 dedication of the monument at Cliffs Shaft Mine Museum. In December 2011, four months after the dedication of the memorial at Cliff's Shaft, Dewey rejoined his father in eternity.

25th anniversary of Marian Tippett Hill and Russell Hill in 1955

Looking back, Dawne (Mrs. Lowell) Smail's only recollection of her daddy was

being on her parents' bed with her twin sister Dixie one morning and sliding down her father's legs. She was only two at the time, and said the girls didn't understand why people brought them presents after the Barnes-Hecker tragedy. They received "duckie dishes," and Dawne's have been handed down in her family.

The surviving Tippett children were actively involved in preservation of the Barnes-Hecker legacy. In 1987, Dawne worked with other descendants, the Ishpeming Rock and Mineral Club, and the Michigan Iron Industry Museum to relocate the granite monument from Ely Township to the grounds of the museum, where more people could view it. In 2001, Irene (Mrs. Bernard) Vicary and Dewey Tippett were interviewed in the WNMU Public TV13 documentary, "Memories of a Misfortune," and the family provided information and photos for the 1998 book, *No Tears in Heaven,* by Thomas Friggens.

Marian Bengry Tippett Hill left a legacy of courage, gratitude and faith – seasoned liberally with good-natured mischief. Walter had had more than one previous brush with death, and she believed he perished in the Barnes-Hecker because it was his time. When she died at home in her 86th year, she departed earth to embrace her Tarzan once again.

–Descendants of Walter and Marian Tippett

See also Chapter 12: Lone survivor; Chapter 20: Catalyst for change; Chapter 11: Timber; Chapter 40: Memory Stone; Chapter 41: Legacy bearers

Dawne Tippett Smail, Dewey B. Tippett and Irene Tippett Lawry Vicary at a family gathering in the 1990s.

Marian Bengry Tippett Hill ca. 1980s. (Photos courtesy of Tippett descendants)

27

Fullback
James (27) and Marie Green (Greene) Trammer

In 1919, James "Bruiser" Green had been a fullback on the Ishpeming-Negaunee Allstars football team. The local team holds the distinction of playing the Green Bay Packers in the Packers' first road game on October 19, 1919, at Union Field located between Ishpeming and Negaunee.

At that time, there were no professional football teams, and the Packers were a "town team," seeking to challenge some of the toughest teams in the central and western Upper Peninsula. The Packers fielded 22 men, while the local all-stars played with 11 men. Despite being outmanned, the local team held the Packers to a score of 33-0, which was the lowest Packers score for the season, and the smallest margin of victory.

Football was a rough sport, and the men who played on local town teams often suffered broken bones, dislocated shoulders and miscellaneous other injuries. After the game, Packers coach Curly Lambeau complained about "dirty tricks" on the part of the all-stars.

At the time of the 1926 Barnes-Hecker disaster, James and Marie Green lived in Ishpeming. He had been employed at Barnes-Hecker for one year, and his body remains entombed in the mine.

According to Cleveland Cliffs Iron Company records, Marie remarried in 1933, when her surname became Gaboury.

James and Marie Green in Milwaukee, 1923 (Courtesy descendants of Walter Tippett)

28   Newlyweds
Arvid (25) and Tyyne (Toini Aalto) Heino
Miner

Born in Finland on Christmas Eve in 1900, John Arvid Heino immigrated to America with his parents and sister Lempi. His sister Helen and brother Arne were born in America. In 1925, Arvid joined his father, Frank Heino, in the workforce at the Barnes-Hecker Mine.

On November 1, 1926, he exchanged wedding vows with Toini Aalto (Tyyne Alto) at the Finnish Lutheran Church in Negaunee. Tynne had also been born in Finland, and immigrated in 1905. Two days after the wedding, Arvid perished when a cave-in occurred at Barnes-Hecker. His body was never recovered.

In its reporting following the tragedy, *The Daily Mining Journal* captured the feelings of Tyyne, who wept as wedding furniture was being delivered while she wore the black attire of a widow. In the years following the tragedy, Tyyne lived with her father, and they took in boarders. One of them was John Hegman, whom she married in 1931. They lived in the neighborhood of West Ishpeming.

John Arvid Heino ca. 1919

Although Arvid had no children, his family members have saved photographs and his Finnish-language Bible. They also saved the letters he wrote to Tyyne during their courtship.

Initial reports of the tragedy had listed both Arvid Heino and his father, Frank, among those who perished, but Frank was unable to report for work that day. Imagine this horrified father, who must have been doubly crushed by the knowledge that he had been spared while his son was lost. And imagine Frank's wife Maria who grieved while trying to ease her husband's emotional pain.

Finnish Language Bible of John Arvid Heino

Paul Heino, grandson of Frank and Maria, believes his grandfather suffered greatly with survivor guilt. In an era when people were expected to be tough, and there were no psychological supports for emotional trauma, Frank Heino sought solace with alcohol. Tragedy compounded tragedy when Frank perished when he was struck by a car in 1934.

Maria Heino ("Mumu" to her grandchildren) lived in the Cleveland Location in the high hills southeast of Ishpeming. She had strong support from family and friends living nearby.

Paul's father, Arne Einar Heino, was the eldest son living at home, and he became the man of the house after Frank died. Arne had to leave school after 7th grade. The family had a small farm, with the barn located about a mile from the Heino home. Arne had to walk there twice a day to care for the cows.

During the great depression, Arne worked for the Civilian Conservation Corps (CCC). As World War II approached, Arne was not accepted for military service, so he served his country by working for a shipbuilding company on the west coast.

Tragedy seemed to follow Arne, who lost his first fiancé to pneumonia in 1936. After the war, he was a jack of all trades who became employed by CCI and other mining companies in Marquette County. He worked at the Humboldt, Maas, Mather B, Bunker Hill, Blueberry, Republic, and Ohio mines. His favorite was the Maas because it was the cleanest and driest.

Arnie married Irene Keskimaki in 1948, and they had two sons, Paul and Jon. Arne and Irene assumed responsibility for the Heino family home on Diorite Street, and Maria lived with them for the rest of her life. She was 85 when she died in 1957. Paul Heino was only three years old when Maria died, and he only remembers that she seemed gray and elderly.

Finnish heritage has become an important aspect of Paul Heino's life. He is taking Finnish Language lessons, and he and his wife hope to travel to Finland.

In the fall of 2016, Paul was among the audience at one of the Barnes-Hecker Remembrance events in observance of the 90th year since the mine disaster. He said it felt somber. He thought about the lives of his dad and siblings as they were growing up. "They were children of the depression who had hard lives," he said. In some ways, the family experiences taught them hard work. "I sure wouldn't have wanted to live through that," he said.

—Jon Heino and Paul Heino

Heino family ca. 1919. Front, Irma Haavisto, Arne Heino, Helen Heino; middle Isaac Haavisto, Lempi Heino Haavisto, Maria Heino; back, Arvid and Frank Heino (Photos courtesy of Paul and Jon Heino)

29 Pandemic survivor
Theodore Kiuru (27)
Miner

Ted "Tego" Kiuru would have been about 18 when he enlisted in the Navy. He was still in training at Great Lakes Naval Training Center (now Recruit Training Command) in Great Lakes, Illinois, when he contracted influenza during the Flu Pandemic that lasted from February 1918 to April 1920. When he recovered, he received an honorable discharge.

Ted helped to support his disabled father, and he was unmarried at the time of the Barnes-Hecker Disaster. He had worked at Barnes-Hecker for three months. Because his job title was "miner," he would have been among

Theodore Kiuru
in Navy uniform

the most experienced and highest-paid men underground. Miners were incentivized with pay based on the *footage* of ore they produced, vs. receiving an hourly rate.

He was a good friend of John Arvid Heino, who also perished in the disaster.

According to the 1927 report from the Pension Department of Cleveland Cliffs Iron Company, there was concern about the welfare of Ted Kiuru's disabled father, John Kiuru, who was the beneficiary of Ted's death benefit. The records show that the company arranged for

Theodore Kiuru

Back row, second from left Theodore Kiuru, third from left (John) Arvid Heino, with friends at Midway Hall on M-35. (Photos courtesy of Margaret Gravedoni)

James Flaa of the American Legion to work on securing "Government Assisted Compensation" for Mr. Kiuru. Those payments began in January of 1928.

Ted Kiuru's great-niece Margaret Gravedoni has been a steward of Ted's history and his tragic loss in the Barnes-Hecker tragedy. Her grandmother, Ethel Kiuru Lawer, provided information for Ted's death certificate. Although Ted left no descendants, he had a girlfriend who surely grieved his loss along with Ted's family.

Margaret Gravedoni said her great-uncle's legacy was one of responsibility and the importance of having good friends.

—Margaret Gravedoni

Echoes
Peter (38) and Evangeline DeRoche
Pumpman

When Peter DeRoche's grandson Bill "Studge" Anderson was a kid in the 1940s, he and his buddies had the freedom to roam the woods around the North Lake neighborhood and westward toward the site of the Barnes-Hecker Mine.

The boys would sometimes squirm through the fence around the abandoned mine shaft, and one of Bill's chums would climb down into the Barnes-Hecker and entertain his friends by making an echo.

On August 31, 2002, echoes of the past resonated in the minds of the families of the 51 lost men of Barnes-Hecker, as they gathered for the dedication of the new monument on the concrete cap of the mine. Among the hushed crowd of more than 100 people were Bill's mother Helen DeRoche Anderson and her brother Donald DeRoche. They were the last surviving children of Peter and Evangeline DeRoche.

Helen DeRoche Anderson and Donald DeRoche at 2002 dedication of Barnes-Hecker monument on the capped shaft.

Peter and Evangeline DeRoche wedding in 1911. Their attendants were Dolfus DeRoche and Louise Neault. (Photos courtesy of Bill Anderson)

Helen later told her son Bill that she had had a life-long recurring dream about her dad. In her dream, he would climb a ladder to her room and watch over her. After that day in the dappled woods, when a bell chimed after each man's name was read and prayers were said, the dream that had echoed through Helen's life for 76 years never happened again. She finally felt closure.

Helen's father, Peter DeRoche, worked in the pumphouse 1000 feet below the surface of Barnes-Hecker. On the morning of November 3, Peter had worked the 12-hour night shift and agreed to work a double shift after another man was unable to report for work.

Peter and his wife Evangeline lived in the Barnes-Hecker Location about a quarter mile north of the mine. Bill Anderson said that until water was plumbed into the homes in 1925, the whole neighborhood was served by hand pumps, one of which was across the street from the duplex where the DeRoche family lived. On warm days, the men would gather at the pump after work to wash up.

Following the disaster, Evangeline had five children to care for, and was expecting her sixth. She worked as a housekeeper for North Lake Mine Superintendent Charles Stakel, and then at the Mather Inn in Ishpeming, where she met and later married George Rock.

Among the De Roche children, Peter worked in the CCI research lab, and later as the caretaker at Cliff's cottage where visiting executives stayed. Helen was married and had a family of 10. George perished in a car accident at age 19. Donald was a full time National Guardsman working in maintenance and strategic planning. Earl, having been imprisoned by the Germans during WWII, was found and released by his brother Don during the occupation of Europe after the war ended. Ethyl, born after the Barnes-Hecker tragedy, was married, had a family and died when she was quite young.

Bill Anderson's son Pete summed up his great grandmother's legacy: "She didn't let the tragedy define her. It made her stronger." Evangeline and Peter DeRoche's legacies are among the echoes that will resonate through scores of their descendants for generations to come.

– William Anderson, Peter Anderson, and Terry Anderson

See also Chapter 51: The Barnes before the Barnes-Hecker

Peter Anderson, who bears the first name of his great grandfather Peter DeRoche, commissioned this painting, "Four generations of Andersons" by Roger Junak.

From left are Peter S. Anderson, his father William R. Anderson, Pete's great grandfather "Captain Ted" Theodore Anderson, and Pete's grandfather William T. Anderson.

31 The deed
Arvid (35) and Sanna Kallio
Miner

About a month before the Barnes-Hecker tragedy, Arvid Kallio made the final payment on a 40-acre parcel of land in Rumley, Michigan. He had already stacked timber there for a farm house. During Arvid's childhood in rural Finland, property ownership was out of reach, and the dream of owning land was a strong lure that drew him to America.

Arvid's wife Sanna had also come from rural Finland, and as a child on a farm, she was tasked with tending the cows. During the summer, she had no shoes, and her feet would bleed from walking on the stiff stubble of cut grass. Sanna came to America alone at the age of 16, and worked for a time as a housekeeper in Chicago.

Arvid and Sanna Kallio

How and when the Kallios met and married is unknown, but they are listed in the 1920 Census living in the Barnes-Hecker Location with their three-month-old son Toivo.

Men with the title of "miner" were paid based on the amount of ore they produced, so when the Barnes-Hecker had to halt mining oper-

Front, Sanna Kallio, Toivo Kallio, Arvid Kallio; back Arvid's sister Tati Lampi and unknown, ca. 1923.

ations due to water problems in 1920, a lot of men left for work at other local mines.

One of those mines was the Salisbury, which closed at the end of June in 1924. When the Salisbury closed, the Barnes-Hecker hired 75 of the men and added a second shift. Arvid is one of several lost Barnes-Hecker men who all have the same hire date of July 7, 1924, so he apparently had worked at Salisbury before returning to the Barnes-Hecker.

By October of 1926, Arvid and Sanna had two children and the future held the promise of dreams fulfilled. Arvid regarded his work in mining as a means to an end. In the fall of 1926, he completed five years of payments totaling $360 for his farmland and had the deed in his possession.

Everything changed on November 3, 1926. Arvid was one of 51 men trapped underground when the Barnes-Hecker Mine caved in. He is one of 41 men who remain entombed there.

Sanna wrote an eloquent Finnish language memorial to

Toivo Kallio, Taimi Kallio Walimaa, children of Arvid and Sanna, ca. 1926

her husband. In it, she said many of the best work heroes had fallen early. She said the tragedy had snatched away her husband and the security of her children. "Why is there in my depressed mind a fire that keeps burning?" she wrote. "I baptize your grave with my tears, other flowers I do not have."

Sanna's grandson David Kallio believes that bitterness and regret may have inhibited his grandmother's ability to recover from the tragedy. She never remarried.

Sanna used her skills as a baker and housekeeper to support her children. In later years, her son Toivo helped to support his mother. Daughter Taimi survived polio. Both of the Kallio children married and had families, and Toivo achieved success as a businessman.

Because Sanna was fluent in Finnish but not English, she couldn't read the text on the deed for the land her husband had paid for, and she would have had to travel to Alger County to have the deed registered. As a result, the deed was never registered and the land was

Toivo Kallio with his bike and Ishpeming H.S. in background

forfeited for unpaid taxes. Grandson Dave Kallio still has the deed, and he has visited the land that slipped through his family's grasp.

At Dave's camp in Marquette County, there is sign that reads "Kallio." It replicates his father Toivo's penmanship. Dave recalls a story from his dad about being a kid in a lumber camp where Arvid and Sanna worked for a time as cooks. "Dad told me that sugar lumps were a little form of candy that he could sneak every now and then," he said. Dave now keeps a bowl of sugar lumps on the table at his camp, and his voice trails off as he says, "Whenever I pop a sugar lump into my mouth, or have coffee with a sugar lump..."

– David Kallio

David Kallio had the penmanship of his father Toivo copied for a sign at David's camp. (All photos courtesy David Kallio)

Sanna and daughter Taimi

32

Boys and grasshoppers
Solomon (49) and Wilhelmina Millimaki (Myllimaki)
Timber hoister

Solomon and Wilhelmina (Minnie) Millimaki were married for 26 years by the time Solomon perished in the Barnes-Hecker tragedy. Both were Finnish immigrants. They lived in the neighborhood of West Ishpeming and had eight sons: Matt (25), Bill (23), John (21), Otto (20), Arne (18), Walfred (15), Swante (14), and Roy (8). Their daughter Minnie was 13.

Bill became a glass blower at a neon sign factory in Waukegan, Illinois, and he made glass peashooters for some of his nephews. Walfred was a carpenter and business agent for the Carpenters and Millwrights Union. Roy owned a heavy equipment company. He had played minor league

Wedding photo of Solomon and Wilhelmina Millimaki

baseball and pitched on an all-star team coached by Babe Ruth. Some of the Millimaki sons worked in mining. Daughter Minnie Millimaki Chipman ran the lunch room at West Ishpeming School, and her cooking was legendary.

Front Minnie Millimaki Chipman, Minnie Millimaki; back Matt "Plush," Roy, Otto "Shiek," Arne "Jack," Walfred "Cush," Bill, Swante "Smoky," and Iohn "Bronco," ca. 1940s

Arne's son Jerry said his dad talked about a time when the family lived on a farm in North Lake. Arne had described Solomon sowing hay seed in the spring by walking through a field with a bag of seed slung around his waist and flinging seeds by hand as he walked along.

Jerry said that after the Barnes-Hecker tragedy, a family member purchased one of the duplexes from the Barnes-Hecker Location for $150, dismantled, and rebuilt it on River Street in West Ishpeming. Jerry remembers watching workmen apply plaster to the lath walls after the house was built.

Roy's daughter Connie Snell said Solomon was known as a rather solemn man with a droll sense of humor. Solomon had joked that the only things he could raise were boys and grasshoppers, but he had a very soft spot for his only daughter Minnie. When Minnie was small, her brothers were tasked with doing her bidding. On one occasion, she wanted to go sledding during the summer, so her brothers had to pull her around the yard on a sled.

Minnie Millimaki Chipman's son Roger said that when he was a kid, he and Grandma Minnie were both early risers, and when she would spot Roger playing outside early in the morning, she'd call him over for homemade pancakes. It was treasured time. Roger said wistfully, "I never got to know my grandpa, and that's something I dearly missed. I wish I could have seen his relationship with his wife and kids."

Millimaki descendants provided fascinating insights into the family's home life in West Ishpeming. Solomon and Minnie's house on the corner of Silver and River Streets had a spring in the basement, and it provided refrigeration for milk and other perishables. Once a week, an Italian grocer from Ishpeming would visit the neighborhood to take orders from the women. Most of the neighborhood families at that time were Finnish, but both nationalities had learned enough of each others' languages to get the correct lists of groceries delivered every week.

In the 1920s, homes in West Ishpeming had outdoor privies, and every spring the contents needed to be dug up and disposed of. Connie Snell said a man with a horse-drawn "poop wagon" would come around every spring to collect the waste and dump it in an area about a half mile away from the homes. The horse had a droopy countenance,

Minnie Millimaki, far left, with Bethel Lutheran Church sewing group. (All photos courtesy Millimaki descendants)

and if kids were cranky, they'd be teased about looking like that droopy old horse from the poop wagon.

Following the Barnes-Hecker tragedy, Minnie Millimaki was one of the widows with older sons who could help to support the family. She, in turn, used some of the compensation funds from CCI to purchase property in West Ishpeming and acreage where some of the sons cut timber. Except for Bill, all of the Millimaki children built homes on the land in West Ishpeming, and they enjoyed close relationships for their entire lives.

Minnie remained active at Bethel Lutheran Church, and neighborhood women would gather at her house to work on quilts and listen to ghost stories on the radio. From time to time, Minnie took in boarders, and that was a catalyst for having an indoor bathroom installed in the house.

The grandchildren said Grandma Minnie was one of the women who never talked about the loss of her husband in the Barnes-Hecker tragedy. Connie Snell thought perhaps her grandmother's silence was the only way she could cope and focus forward. As Connie reflected on the photo of Minnie with her adult children, she said, "Grandma looks so content, and I thought for her at that point in time, life was pretty doggone good."

– Roger Chipman, Gerald Millimaki, and Connie Millimaki Snell

Baby Chicks

33

Baby Chicks
Uno (22) and Vieno Koskinen
Miner

Uno Koskinen was one of the 51 men who perished in the Barnes-Hecker tragedy of November 3, 1926. His widow, Vieno, was left to face a daunting future. She and Uno had been married in January, and they were expecting the birth of her first child in the upcoming spring.

At the time of the tragedy, the couple were living with Vieno's parents, Charles and Amalia Jarvi. Vieno's parents comforted her through her grief and the birth of her son Stanley in April of 1927.

In 1930, Charles Jarvi sought a better future for himself, his wife, daughter and grandson, and made arrangements to move the family to New York State. There, they stayed with another Finnish

Uno and Vieno Koskinen

family while Charles built a home and started a poultry farm. The farm was in Trumansburg, New York, about 10 miles from Ithaca, where Vieno found work as a housekeeper in a fraternity house at Cornell

University. Stanley remained at the farm with his grandparents, who took him to visit his mom on weekends whenever possible.

Over the years, young Stanley grew up helping his grandfather while learning the business of operating the chicken farm. Stanley broadened his knowledge by working at the Babcock Chicken Hatchery, and when he became owner of the family farm, he expanded operations to include a year-round hatchery. During the summer, he often took his teen-aged children along when he called on local farmers to secure orders and deliver baby chicks.

The four Koskinen kids— Roger, Karen, Brenda, and Judith—each had assigned

Vieno Koskinen and Stanley Koskinen (Courtesy Koskinen descendants)

tasks to remove newly hatched chicks from the incubators, pack the special cardboard shipping crates, and accompany their dad as he delivered chick boxes to both farmers and the post office for shipping. The Koskinen kids said their dad taught them the value of hard work.

Stanley Uno Koskinen, the only child of Uno and Vieno Koskinen, was raised by his maternal grandparents, and the family story illustrates the critical role played by grandparents in the lives of many of the children who lost their dads in the Barnes-Hecker tragedy. Stanley raised chickens until 1984, and continued growing corn, soybeans and wheat until he was 86 years old.

Stanley's daughters, Karen Koskinen of upstate New York and Brenda Koskinen of Boston, pondered how different their lives would have been if not for the move to New York, in close proximity to a university town.

Karen and Brenda remember their grandmother Vieno as a person who loved to hug and laugh, and was very upbeat and chatty. She remarried later in life.

The Koskinens cherish the legacies handed down in their family. "You have to build a new life." "Whatever dad did, he took responsibility for getting it done." "Whatever you do, do it right." Stanley Koskinen never knew his father. And he felt that loss with an intensified sense of longing as he neared the sunset of his life. At times, it brought him to tears. But the perpetual hatching of downy baby chicks embodied his greatest legacy that with the sunrise, opportunity and renewal would always come again.

—Karen Koskinen, Brenda Koskinen, and Erika Reed

Uno and Vieno Koskinen (courtesy Karen Koskinen)

34 Recipe for Love
Emil (38) and Lempi Maki
Miner

In the latter years of Lempi Maki's life, Saturday mornings meant pancakes for her kids and grandkids. Her small kitchen would often overflow as she made one pancake at a time in a small cast iron frying pan. The memories of those times are precious to her grandchildren, and they didn't realize that a secret ingredient was written between the lines of Lempi's cherished recipe. As family members grabbed a seat and squeezed in wherever there was room, conversation and laughter flowed as each person waited their turn for a fresh hot pancake. On pancake days, in the midst of the happy din, Lempi also fed her families' hearts.

At noontime on November 3, 1926, Lempi had probably been busy in her kitchen making lunch and tending to two-year old Ruth when tragedy came crashing into her life. Lempi's husband Emil was among the lost men of Barnes-Hecker, and he would never be coming home.

Over the years, Emil Maki had honed his mining skills and worked his way up to the job title of "miner." Miners supervised small crews of two or three other men in assigned work areas called "contracts." Rather than earning a set daily wage, contract miners were paid by the "footage"/tonnage of ore they produced, and they were the highest-paid men on the payroll.

Because Emil's hire date at Barnes-Hecker was July 7, 1924, it's probable that he had worked at the Salisbury Mine in Ishpeming, which closed on June 30, 1924. He was one of 75 men hired at Barnes-Hecker when the Salisbury closed. Barnes-Hecker was about seven miles from the city of Ishpeming, and for each shift, the mine would send a large truck with a canvas-covered back to pick up the men who lived in town. Emil missed the truck on the morning of November 3,

Lempi Maki with children ca. 1927. Front, Richard (7), John (8), Ralph (11); back Ruth (2), Lempi, Wilho (12).

and drove the family car to the mine.

Lempi didn't talk about her feelings after the disaster, but over her lifetime she left a powerful legacy. With five kids to support, she went to work at the Gossard women's undergarment factory, where she became a supervisor. She tasked her eldest son Wilho with the responsibility of watching his younger siblings, and because of that, Wilho had to drop out of school. At the time of the disaster the Maki children ranged in age from 12 to 2.

Lempi Maki

Back: Wilho, Ralph, John, Richard;
Front: Lempi and Ruth.

In adulthood, Emil and Lempi's sons Wilho and Ralph worked in mining, and John owned a company that installed and serviced gas stations. Richard earned an engineering degree from Michigan Tech and worked for Jones & Laughlin before leaving the

> **Lempi's pancake recipe:**
> - 1 C. flour
> - 1 egg
> - 1 tsp. sugar
>
> Mix and add milk till sloppy.

area. Ruth became a cosmetologist in Chicago. Lempi was an aspiring artist, and served in leadership roles at her church. She sang Finnish songs to her grandchildren, and after living a very simple life, she left small bequests to each of her five children. She passed away in 1982 at the age of 87.

Lempi's granddaughter Carole Maki Bartanen shared Lempi's pancake recipe, and it's included here with the hope that everyone will find that secret ingredient written between the lines.

–Carole Maki Bartanen and Bruce Bartanen

–Shellly Maki Isherwood

Original oil painting by Lempi Maki

In this undated photo, Emil Maki is holding the accordion-like instrument, and Lempi is behind him holding flowers. Archivists at the Finnish-American Heritage Center (Formerly Finlandia University) said this photo was probably done by an itinerant photographer, and may have been done on a Sunday because the people are wearing their *pyhävaatteet* (Sunday's best). One of the archivists called the instrument in the middle a *pirunkeuhko* (a devil's lung). He said the church viewed the decorated instrument as "too worldly" and unfit for sacred music. Instead, it was suitable for a dance hall.

According to the archivists, the stringed instrument on the right is held like a Finnish kantele, but the angle of the string pegs suggests that it was more likely to be a zither.

Based on Lempi Maki's obituary and the age of the oldest Maki child at the time of the tragedy, Lempi would have been younger than 18 when this photo was taken. If that is true, then the photo was taken before 1913.

All photos in this chapter courtesy of Maki descendants.

Chopping wood
Elias (42) and Amanda Ranta
Miner

When Amanda and Elias Ranta departed Finland for North America, they had to leave their first two children in the care of family members.

By 1926, Elias had become an experienced contract miner who had worked at the Barnes-Hecker Mine for four years. The couple lived in North Lake, a community about two miles east of the Barnes-Hecker. In 1926, the Rantas had eight children, with 6 under the age of 16.

In October of 1926, Amanda and Elias almost lost their nine-year-old son, Reino, in a horrific accident. Reino remained in the hospital with severe burns and other injuries that cost him his right arm. On November 3, Amanda suffered another overwhelming shock when she

Amanda and Elias Ranta

Reino Ranta as a child   Elias Ranta

received word that her husband had been killed in a cave-in at the mine. She was pregnant with the couple's ninth child.

Amanda's life after the tragedy is embodied by the Finnish word "sisu" which translates as a combination of fortitude, tenacity, grit, stoicism, strength, and grace under pressure. She made smart choices and used a portion of the CCI compensation funds to purchase a home on Cedar Street in Ishpeming. At some point, Amanda went to work in a mattress factory. The only known mattress factory in the region, still in operation in 2025, is near the town of Escanaba, Michigan, about 90 miles from Ishpeming. It opened in 1936. It's likely that Amanda would have had to travel there by train and stay in the Escanaba area during the workweek.

When Reino Ranta came of age to apply for a job, he had difficulty being hired because he had only one arm. He sought advice from his mom, who said, "Why don't you show the foreman how you can chop wood?" And so he did. Reino ultimately finished college and became a

Amanda Ranta with seven of her nine children, ca 1927 (All photos courtesy of Ranta descendants)

long-time high school teacher and debate coach in Baraga, Michigan. He then taught for 21 years at Suomi College, which became Finlandia University in Hancock, Michigan.

Reino's daughter Elaine Ranta Schindler said her dad took pride in doing physical things just as well as anyone with two arms. He crafted his own large snow scoop, and in a region that typically has more than 200 inches of snowfall each winter, it's a daunting task for *anyone* to clean up snow by hand. For as long as he was able, Reino enjoyed chopping wood, and taught one of his daughters how to do it.

For Reino Ranta's entire life, he chopped down barriers, and his mother's words, "Why don't you show them how you can chop wood?" became a metaphor for overcoming challenges and attaining achievement.

–Elaine Ranta Schindler

36

The Other Side

George Lampshire (34) and his wife Alice
Richard Lampshire (23) and his late wife Lena (née Cowling)
Earl (28) and Sarah Lampshire Ellersick

George and Richard Lampshire were brothers and miners who had given notice that they had accepted other jobs. November 3, 1926, was to be their last day at the Barnes-Hecker Mine. It was the cruelest of ironies that they never came home.

George was married, and he and his wife Alice had a two-year-old daughter. Richard was a widower whose wife, Lena Cowling Lampshire, had died of typhoid in 1925. Richard and Lena had four children who were orphaned by the Barnes-Hecker tragedy.

The Lampshires' sister, Sarah Lampshire Ellersick, lost two of her brothers and her husband, Earl Ellersick, who was a trammer. Sarah and Earl had four children.

Reporting since 1926 has stated that the orphaned Lampshire children were raised by Sarah and her mother, Mary Ann Lampshire. CCI Pension Department records tell another story. The records reveal that the children were raised by their maternal grandparents, Esther and William Cowling.

In another twist, the Cowling family was a link to Thomas Kirby Sr. and Thomas Kirby Jr. One of the Kirby daughters, Rita, was married to Charles Cowling, who was a brother to the late Lena Cowling Lampshire. Collective grief was thus shared by the Lampshires, Ellersicks, Kirbys and Cowlings for the loss of five men. Rita Kirby Cowling's granddaughter, Georgann Kippola Coon, confirmed that Richard Lampshire's children were, indeed, raised by their maternal grandparents, Esther and William Cowling.

For many years, Lampshire-Ellersick descendant Jack Perala of Negaunee has sought out information about his lost great-uncles

Memorial tribute to Richard and George Lampshire (photos courtesy Gary and Ronald Perala)

George and Richard and the tragedy that claimed their lives. Jack worked at the Morris-Lloyd Mine in the 1950s. The Morris connected through an underground tunnel with the Barnes-Hecker, and a dam had been erected there in 1926 to prevent water from entering the Morris. Jack became part of a crew assigned to construct a second dam when the older one began to leak. In a 2016 interview with *The Mining Journal*, Jack said, "I had the privilege of going up to the old dam and putting my hands on there. And my uncles were on the other side," he said.

– Ronald (Jack) Perala, Gary Perala, Georgann Kippola Coon, and Craig Cowling

See also Chapter 16: Pocket watch

Ronald (Jack) Perala (age 89)
Courtesy of Gary Perala

37

No longer anonymous
Olaf Trondson (1897 - 1979)

In 2016, the Barnes-Hecker Remembrance posted the well-known photo of men posed around a large piece of equipment that resembles the front of a locomotive. Kate Trebilcock Sromalski posted a Facebook comment that the man standing on the left, leaning on the device, was her grandfather, Olaf Trondson. She said the ID had been

Olaf Trondson (Courtesy of Michigan Iron Industry Museum)

done by her uncle, Trondson's son, Paul. She also referenced her grandfather's military record, which shows his World War I service in the Army from July 11, 1918 until October 10, 1919. This record narrows the time period of Trondson's employment at Barnes-Hecker from October 1919 until Trondson left the employ of Barnes-Hecker in late 1920.

In the Barnes-Hecker Annual Report for 1920, there is the first mention of horizontal drifting (tunneling) to open the three main levels at 600, 800 and 1000 feet below the surface. That report also states that on September 9, 1920, heavy flows of water occurred, and work was temporarily halted. Many of the men left the employ of the mine at that time. Unfortunately, there are no payroll records prior to 1921.

Olaf Trondson's obituary notes that he was a World War I veteran, and worked for Bell Telephone Company for many years. He was survived by three daughters, two sons, five siblings, and numerous grandchildren and great-grandchildren.

– Kate Trebilcock Sromalski

See also: Chapter 49 Sub-level caving by the top-slicing method

Roots and branches

38

Edwin (46) and Elsie Chapman
Miner
Herman Chapman (22)
Trammer

When the Barnes-Hecker went down on November 3, 1926, Elsie Scantlebury Chapman was one of the widows who grieved the loss of both her husband Edwin and son Herman. Neither of the Chapmans were recovered.

The Chapman family roots were in Cornwall, England, where Edwin and Elsie wed in 1901. The first three of their ten children were born in Cornwall.

The family immigrated prior to 1908 when their fourth child was born. They lived on Excelsior Street in the Salisbury Location south of Ishpeming. Based on the July 7, 1924, hire date for both Edwin and Herman Chapman, it's probable that they both worked at the Salisbury Mine until it closed at the end of June in 1924. Following the Barnes-Hecker tragedy, Elsie bade farewell to her 16-year-old son Gordon in late November when he left for Shamokin, Pennsylvania with his aunt and uncle, Mr. and Mrs. James Chapman. The eldest daughter, Doris Mae Chapman Antilla, died in 1927 in Ishpeming. Two of the oldest Chapman children lived in the Detroit area at the time of the Barnes-Hecker tragedy, and at some point Elsie also moved there with the youngest children.

Elsie's daughter Nora, the eighth of the ten Chapman children, was nine years old when the Barnes-Hecker tragedy happened. As an adult, she worked for Chrysler in Detroit. She was in her thirties when she married James Crass, who was a machinist at Chrysler.

James Crass and his first wife had moved from Kentucky to Detroit after World War II. With few jobs available in Kentucky, he accepted a

This photo of the Chapman family ran in *The Daily Mining Journal* in November of 1926. Edwin Chapman is seated on the right, and Herman is seated on the left. Standing from left are Chapman daughters Mrs. Doris (Evar/Eavar/Ivar) Antilla, Mrs. Gladys (Sidney) Pengilly of Detroit, Vera (12), Nora Ivy aka "Joy" or "Peggy (9)," and sons Cecil (18) of Detroit, Gordon (16), and Alvin (14). Seated next to Herman is his mother Elsie and four-year-old Ellen. Six-year-old Douglas is leaning into his Dad's lap. Doris Chapman Antilla died in 1927; Alvin Chapman died in 1943; both are buried in the Ishpeming cemetery. (Courtesy of Mike Draffen)

job at Chrysler and they lived in an area known as "Little Kentucky." A daughter, Janice, was born in Hamtramck in 1946. After James's wife deserted the family, he married Nora Chapman in 1952. Nora became mom to Janice, and Nora and James adopted Donald and Mary who were biological siblings. James inherited the family farm in Kentucky, and the family moved there.

James had been 18 when he was rejected for military service during War II, so he worked as a machinist in a factory in Paducah, Kentucky. When the plant closed, he was transferred to Oak Ridge, Tennessee, where he had to live onsite at the factory. After the war, he wasn't permitted to leave the country for many years but his family didn't know why. When he was allowed to disclose the reason, it was because the Oak Ridge factory made components for the atomic bombs that were dropped on Hiroshima and Nagasaki at the end of World War II.

Chapman family home village of Cardinham Parish, Cornwall, England. (Courtesy of Mike Draffen.)

As the years passed and the children married, Nora spent much of her life alone and unhappy on the farm while James was away working as a machinist. Nora's son-in-law Robert Draffen said there was little choice in the matter, because jobs were few, and people either worked their farms or starved. A machinist could provide a higher standard of living for his family.

Through the years, Nora kept family keepsakes and photographs of the Chapman family and their ancestral home village of Cardinham in Cornwall.

AlthoughNora's children and grandchildren were not related to her by blood, they knew she had lost her father and brother in a mine accident in Ishpeming, Michigan. Nora's grandson Mike Draffen found Ishpeming online and that's when he learned about the Barnes-Hecker tragedy. Mike had become the keeper of Nora's keepsakes, and he wanted to make sure the photos and documents would be available to other Chapman descendants.

Robert and Mike suffered a loss of their own when their beloved wife and mom Janice Crass Draffen passed away in 2018. Janice was about six years old when Nora became her second mom. Janice became

This photo has no identification, and is Chapman and/or Scantlebury ancestors in Cornwall. The style of the men's clothing suggests the photo was taken in the late 1800s or early 1900s. Edwin and Elsie Scantlebury Chapman were married in 1901 in Cardinham. (Courtesy of Mike Draffen.)

a nurse, and when Nora was in the end stage of breast cancer, Janice arranged to have Nora on the same hospital floor where Janice was the charge nurse. She was at Nora's bedside when she passed away in 1995.

Robert said he had barely known Nora's elderly mother Elsie Chapman who was a very quiet person. He said that both Elsie and Nora made Cornish pasties (pronounced with a short a) that are still made by descendants in their family. Pasties are meat pies with sliced potatoes, rutabaga, and onions, seasoned with salt and pepper. When served family-style, pasties are made like pies. Individual pasties have a generous helping of ingredients wrapped in a crimped crust.

The history of pasties is closely tied to mining. In the earlier days of mining, a miner's lunch pail had compartments for hot coffee in the bottom, a pasty in the middle, and a sweet item on top. Underground miners had no hand-washing facilities, so they could hold a pasty by the crimped edge of the crust and discard the crimp after their meal.

Robert said his mother-in-law Nora Chapman Crass, whom the family called, "Peggy," was a good mother to her children, with ups and downs typical of any family. Life had not been easy for her, but she had a loving family who took care of her as she had cared for them.

From the Chapman family roots in Cornwall to a branch of the family tree in Kentucky, the Draffens have added a fascinating dimension to the broad legacy of the Barnes-Hecker tragedy.

—Robert Draffen and Mike Draffen

This booklet tells the 1905 story of the Chapman family. (Courtesy of Mike Draffen.)

39 Everyone goes to their destiny
Ed Temo – Erkki (48) and Amalia Timoharju Miner

In 1898, Erkki and Amalia Timoharju were married in Lappajärvi, Finland, where Erkki inherited the family farm. The couple suffered the loss of their first child, Tuovi Inkeri shortly after birth in 1899. Two daughters, Hilma Maria and Eveliina were born in 1900 and 1903.

Many Finns were emigrating to America at that time, and Erkki felt the lure of adventure and opportunity in the copper mines of Michigan's Upper Peninsula. He found work in Calumet, not far from his sister Maria Korri and her family in Houghton.

After two years, Amalia entrusted her girls to their grandparents and joined her husband in America. In Calumet, she liked the ease of having milk delivered every morning, and being able to shop for anything the couple needed.

Erkki and Amalia Timoharju in America in 1907. (Courtesy Timoharju

Timoharju children at the family farm in Lappajarvi, Finland, ca. 1923. Back, Juho Erkki, Oskari. Middle, standing from left Helviina and Anna. Front Eveliina (seated) with Saima on her lap, Martta, Erkki Timoharju's sister Reeta Orava, Anselmi. The youngest child, Viljo, wasn't born until 1925 after Erkki left to return to America. (Courtesy Timoharju descendants.)

While in Calumet, the couple had a son, George, who died shortly after birth in 1907. Another son, Juho, was born in Calumet in 1908. By 1909, the couple felt they had made a mistake by coming to America. Separation from the girls had also become increasingly difficult for everyone, so they boarded a ship bound for their homeland. Their son Oskari was born aboard ship in the North Sea.

After they returned to Finland, they had seven more children between 1910 and 1925: Selma, Matti, Helviina, Anna, Anselmi, Martta, Saima and Viljo. During the years on the farm, expenses grew, and the couple realized it was a mistake to leave America. Even though Amalia was pregnant for Viljo, Erkki left again in December of 1924.

The Timoharju family farm in Lappajarvi, Finland, is now owned by granddaughter Suvi Nyyssölä and her husband Jorma.

It was a difficult decision. When he arrived in America, he was hired at the Barnes-Hecker Mine, and was one of the 51 men who perished in the cave-in of November 3, 1926.

Amalia's older brother John Erick Söyring (Juho Erkki Söyrinki) lived in Gwinn, Michigan, about 30 miles from Ishpeming. He was employed at the Princeton Mine owned by Cleveland Cliffs Iron Company. He had the heart-wrenching task of contacting his sister with the news that her husband was one of the men killed in a cave-in. His youngest son later said it was the most difficult thing he had ever had to do.

Söyring provided information for his brother-in-law's death certificate and helped Amalia with the paperwork necessary to receive compensation that was provided to all of the widows. Because CCI required proof that the lost men were sending support to Finland, Amalia had to provide them with all of her husband's letters. She never got the letters back.

At the time of the tragedy, there were 10 surviving children ranging in age from 26 to 1. The compensation from CCI made it possible for

Amalia to keep the younger children with her, buy more land, and allow Anselmi to remain in school instead of staying home to help with the farm. In 1931 she lost her youngest son Viljo in a tragic accident.

Daughter Helviina was 13 when she lost her dad, and she was angry that her father had left her mother alone to raise the children and manage the farm. The boys lost a fatherly role model to teach them how to farm, fish, ride a bike or care for the farm. Helviina later said that she had fond memories of Erkki playing and singing with his children, who had inherited his talent for music.

John Erick Soyring with grandson John A. Soyring, 1954. John Erick lived in Gwinn, Michigan, and helped his sister, Amalia Timoharju in Finland, with paperwork needed to receive her husband's death benefits. (Courtesy John A. Soyring)

Amalia Timoharju kept her feelings about the tragedy inside. She never remarried. Raili Rönkkö said her grandmother lived meaningfully in the conditions she was given. She remembers her grandmother as a playful and nice person who exemplified the Finnish word *sisu*: a combination of strength, stoicism, determination, and grace under pressure. She said one of her grandmother's favorite sayings was, "Everyone just goes to their destiny."

– Suvi Nyyssölä, Raili Rönkkö, and John A. Söyring

Memory stone
The granite monument from Ely Township-1971

On August 27, 1972, a crowd gathered for the dedication of the first monument to the men who perished in the Barnes-Hecker tragedy. The granite stone stands six-and-a-half feet tall, weighs almost two tons, and is engraved with the names of the 51 men who perished in the Barnes-Hecker Mine Disaster.

The monument is a tribute commissioned by the citizens of Ely Township during their 1971 centennial celebration.

At the time the monument was created, the private ownership of the Barnes-Hecker land posed a challenge as to where the monument should be placed. The initial solution was to place the monument near the former Evergreen Drive-in Theatre on U.S. 41, about seven miles west of Ishpeming.

Former North Lake Mine Superintendent Charles J. Stakel with lone survivor Rutherford Wills at the 1972 dedication of the monument commissioned by the citizens of Ely Township.

Lillie Tippett LaBeau, Rutherford Wills, Jennie (Jean) Tippett Wallberg, Evangeline DeRoche (Peter DeRoche), Nellie Tippett (Captain Bill), Marian T. Hill (Walter Tippett.)

The 1972 dedication, on the original site of the monument, was attended by former North Lake Mine Superintendent Charles Stakel; Rutherford Wills, the only man to escape the flooding mine; Ely Township dignitaries; as well as widows, children and grandchildren of the lost men.

In the 1980s, Saima Timoharju Nyyssölä traveled from Finland to visit the final resting place of her father, Erkki Timoharju (Ed Temo), whose name is engraved on the monument. Saima called it a "memory stone."

Saima's words capture the emotional impact of the gift given to posterity by the citizens of Ely Township. Gratitude for that gift transcends the boundaries of distance and time.

In 1988, after the Michigan Iron Industry Museum opened in Negaunee, Michigan, a group of descendants worked with Ely Township, the museum, and the Ishpeming Rock and Mineral Club to move the monument to the grounds of the museum where more people could view it and pay their respects to the lost men.

Marian T. Hill and Nellie Tippett lost their husbands, Walter Tippett and his brother Captain Bill Tippett.
(Dedication photos courtesy Michigan Iron Industry Museum.)

Clifford Trudell, Sr., points to the name of his lost father, Louis Trudell.
(Courtesy of Clifford Trudell.)

Saima Timoharju Nyyssölä at the original site of the Ely Township
monument ca. 1980s. (Courtesy of Suvi Nyyssölä)

41 Legacy bearers
Ely Township Monument rededication

It was Thursday, September 15, 1988, when a flatbed truck rolled into the lower parking lot at the Michigan Iron Industry Museum. Precious cargo was swaddled in padding and strapped securely in place.

A group of people stood waiting to greet the truck and its cargo. The assembly included daughters and sons of the men who had worked at the Barnes-Hecker Mine. They were waiting to witness the arrival of the 1900-pound Ely Township monument to the 51 men lost in the tragedy of November 3, 1926. The monument had arrived at its final resting place at the Michigan Iron Industry Museum.

When the monument was originally dedicated in 1972, the Iron Industry Museum didn't exist.

Because the site of the Barnes-Hecker was and is on private property, the stone was originally placed on U.S. 41 near the Evergreen Theatre about seven miles west of Ishpeming. The grass had grown tall, new generations were born, and for some, the monument was just another relic on the side of the road. . . but not for the people who were children when they lost beloved family members, and not for the people of Ely Township.

For the descendants and the leaders of Ely Township, the monument was at risk of declining into obscurity. After the Iron Industry Museum was built in 1987, there was a possibility for the monument to find a new home on the museum grounds. A purpose-driven collaboration by descendants, Ely Township, Thomas Friggens of the Iron Industry Museum, and the Ishpeming Rock and Mineral Club resulted in applause when that flatbed truck rolled into the museum parking lot. A few tears rolled down the cheeks of the people waiting there.

Thomas Friggens, historian of the Michigan Iron Industry Museum, welcomes the audience. Seated on his left are descendant Dawne Tippett Smail, Ely Township Centennial committee member Jean Warlin, The Right Reverend Monseigneur David Spelgatti, and The Reverend Rudolph Kemppainin.

At the rededication, the crowd overflowed from the sunny parking lot into the cool shade that surrounds the monument.

Many families shot photos with the monument that day, and among them were descendants of William, Walter, and Albert Tippett. From left are Dawne Tippett Smail (daughter of Walter), Patricia Procunier (daughter of Albert), Eric William Tippett pointing to the name of his great grandfather Captain William Tippett, Margaret Tippett Sherwood (daughter of William) with hands folded and family members behind her, Irene Tippett Lawry Vicary (daughter of Walter), unknown, Dewey Tippett (son of Walter).
(All photos courtesy of M. Tippett.)

The re-dedication event on June 11, 1989, was attended by a large crowd of several generations of descendants. During the ceremony, the name of each man was read, and a portion of a 1926 funeral sermon by the Reverend Hugo Hillila was delivered by Rev. Rudy Kemppainen. That sermon included text from Revelation 21:4, "God shall wipe away all the tears from their eyes. There shall be no tears in heaven."

One of the speakers at the event was Jean Warlin, who served on the Ely Township Centennial Committee and was a teacher at Diorite School. She was a member of the steering committee for the monument fundraising effort. Jean said that her grandfather had been unable to report for work at Barnes-Hecker on November 3, 1926. For his lifetime, he bore the emotional burden of survivor guilt because another man had died in his stead. Jean's father, Charles Johnson, was 15 at the time of the disaster and went to the mine site on the day of the tragedy. In her remarks, Jean told her family's story as an example and said, "Please keep passing these stories on! They are a treasure!"

The people who made it possible for the Ely Township "memory stone" to find a permanent home are among the scores of legacy bearers who have stewarded the Barnes-Hecker story for almost 100 years.

PROGRAM SCHEDULE

*   *   *

Barnes-Hecker Mine

MEMORIAL REDEDICATION

JUNE 11, 1989

Monument Rededication	1:00	Parking Area
"The Whole World Can't Stop Because of Our Sorrow" (National Mine School Multi-Projector Slide Show)	2:00; 3:30	Auditorium
History in Music (Mark Mitchell, Balladeer)	2:30; 4:00	Auditorium
Museum Open House	9:30 – 5:00	

*   *   *

The museum gratefully acknowledges the Lake Superior [illegible] House Dineen's and Terry's Central Florist [illegible].

The Michigan Iron Industry Museum is administered by the Bureau of History, Michigan Department of State.

Michigan Iron Industry Museum

Negaunee, Michigan

42   A rock of strength
Nicola (32) and Mary Valenti
Miner

Nicola Valenti and Mary (Maria Donato) were a gorgeous young couple. Mary's wedding gown was inset with lace. Her veil was like a cloud framing her face. The groom was strikingly handsome.

Their marriage license, dated May 17, 1913, stated that she was 18 and he 21. In truth, Mary turned 15 two months after she married Nick, who was 18. It was an arranged marriage. The ages listed on the marriage license were provided by the couple's parents.

By 1926, the Valentis had seven children, and Nick had a good job in mining. On November 3, 1926, the world crashed in on Mary Valenti when word came that her husband was among the men likely to have perished in a cave-in at the Barnes-Hecker Mine. Nick had not been scheduled to work, but because his brother Anthony was ill, he agreed to work in his brother's place. Nick's body was never recovered.

The wedding of Nicola and Mary (Donato) Valenti. Their attendants were Antonio and his wife Lisa Betha Carello.

Nick Valenti with his father Frank and brother Anthony. Anthony could not report for work on November 3, and he became a beloved father figure for the Valenti children and grandchildren.

Mary had given birth to baby Nicholas only four months before the tragedy. Her other children were Frank, 12; Rocco, 10; Anthony, 6; Salvatore, 5; Columbus, 3; and Rose, 2. Although she had help from the extended family living close by, she was hard pressed to manage the household and provide for the children. And she had not had the traditions and rituals of her faith to feel assured that her husband's soul had passed into eternal life.

All of the Barnes-Hecker widows received compensation from CCI, amounting to $28 per week for five years; however, Mary later told her adult children that she had literally begged at local grocery stores and restaurants for food that would otherwise be discarded. Mary's boys did odd jobs to help support the family. Her children would not have had Christmas gifts without the help of the Salvation Army. Mary was a wise woman who required all of her children to speak English. Although she had only a sixth-grade education, she spoke both Italian and English. Her grandson Lee Warner said that she would translate letters for neighbors and friends. When he visited his grandmother, she would ask him to read letters aloud to her in English while she was on the phone translating into Italian.

When Mary was 32, she married 36-year-old farrier Theodore Keskey on June 11, 1930. They enjoyed a 32-year marriage. They're pictured on their 25th anniversary. The Keskeys had two more boys, Theodore Jr., and Frederick.

Theodore and Mary (Valenti) Keskey celebrated their 25th anniversary in 1955

Mary and Ted Keskey with their sons Frederick and Theodore ca. 1940s

In later years, Mary told her children that when she met Ted, she found love. Her marriage to Nick Valenti had not been an easy one. At the time of the 1913 wedding, Nick had longed to return to Italy.

Diana Warner Bushong said her grandmother was never bitter about the way her young life had unfolded. She called her grandmother a rock of strength who taught the family how to love.

An enduring symbol of that love is the obelisk monument that now stands atop the capped shaft of the Barnes-Hecker Mine.

–Diana (Warner) Bushong, Lloyd Bushong, Lee Warner,
Bob Warner, and Mary Warner Hooper

Nicola Joseph Valenti

Photo (left): Valenti & Keskey Brothers. Front Row: Anthony Valenti, Frank Valenti, and Rocco Valenti. Back Row: Salvatore (Tudy) Valenti, Frederick Keskey, Nicholas Valenti, Theodore (Babe) Keskey Jr. and Columbus Valenti.

Source: Genevieve Valenti February 2008

Valenti and Keskey sons, ca. 1950s.

The Obelisk
Monument at the capped shaft
Nicola and Mary Valenti

An early view of the Barnes-Hecker headframe facing west ca. 1918. (Courtesy of the Greg Montgomery/Allan Koski collection.)

Until 1952, the headframe of the Barnes-Hecker Mine stood as a silent sentinel that marked the site of the 1926 tragedy. In 1952, a national shortage of steel led to collection of scrap steel from farms and idled mines and factories around the country. The Ishpeming *Iron Ore* reported that because the headframe could yield an estimated 75 tons of steel, Inland Steel Company decided to have the headframe dismantled for re-use by steel mills.

Ever since the tragedy in 1926, the absence of a memorial marker at the mine site has left a painful void for many families. At a 2001 family reunion of the descendants of Nick and Mary Valenti, a discussion of the unmarked mine site led to a decision: to find out what it would take to obtain permission for placement of a monument on the capped shaft. Members of the extended family from around the country contributed all of the funds needed to create such a monument.

The result was a forged obelisk with a plaque listing the names of the 51 lost men. This chapter focuses on the obelisk, its creation and the 2002 dedication ceremony.

First and foremost, the monument's genesis arose from a family's love for their mother and grandmother, Rose Valenti Warner, who was two years old when she lost her father. Rose's devotion to her Roman Catholic faith gave her a lingering concern about the rituals and traditions that were denied to her father at the time of his passing. There were no last rites, no Italian neighbors to come and chant while a casket would lie in repose in the home, no funeral, and no interment or cemetery marker. No place to leave flowers and recite the Rosary in remembrance. Nothing to mark her father's final resting place.

As ideas for the monument were explored, Rose's son-in-law Bob Hooper, an avid history buff, researched the architecture of Marquette County mines and recommended an obelisk inspired by the A and B shafts at Cliffs Shaft Mine Museum. The family worked with the Michigan Iron Industry Museum, mining historian Leo LaFond, Cliffs Shaft Mine Museum, and the owners of the mine site for permission to place the monument and plan for a dedication ceremony.

Rose Valenti Warner (Courtesy of Valenti descendants.)

Son-in-law Lloyd Bushong worked in the Marquette office of a Minnesota-based company (The Jamar Company), which procured and donated COR-TEN™ steel, which is rust-resistant and acquires a protective patina as time passes. The company also permitted Lloyd and Diana Bushong's son, Beau, who was a welder at the company, to cut and weld the steel on paid time.

Bob Hooper scoured antique shops for a large bell, and when he explained what he was looking for and why, a Negaunee antique shop offered to loan a huge bell for the dedication ceremony.

When the obelisk was completed and plans were in place to hold a public dedication over Labor Day Weekend in 2002, Bushong's company provided use of a truck to transport the monument and the bell to the site of the Barnes-Hecker Mine.

Lloyd and Beau Bushong were alone in the silent woods when they unloaded the truck. They affixed the monument on the concrete cap of the mine and set up the bell, which was mounted on a large frame. It was an experience that brought both men to tears.

On the day of the dedication, many families laid flowers at the base of the obelisk on the capped shaft of the Barnes-Hecker Mine. The mine site is on private property. (M. Tippett photo Sept. 1, 2002)

As each name was read, Beau Bushong, great grandson of Nick and Mary Valenti, pulled the rope on the heavy bell, with the chimes reverberating through the forest 51 times. (*The Mining Journal*, Marquette MI, Sept. 1, 2002)

Diana Warner Bushong and her brothers Bob and Lee Warner said the family thought perhaps 30 people might come to the dedication ceremony. But as the time approached, a caravan of cars made its way to the mine site, and almost 200 people were in attendance. They brought flowers and cameras and gratitude. The first generation of descendants had their children and grandchildren with them. Descendants from several states were in attendance.

Mary Warner Hooper (at the podium) made remarks at the dedication ceremony on behalf of the family. As the name of each lost man was read, Beau Bushong pulled the rope on the heavy bell. As the chimes reverberated through the forest, tears ran down the faces of the audience.

After the ceremony was completed, the Valenti descendants hosted a pasty meal and social time at Cliffs Shaft Mine Museum.

Fifty years after the removal of the Barnes-Hecker headframe, the Valenti descendants' gift to Rose Valenti Warner became a gift for the ages.

– Diana (Warner) and Lloyd Bushong,
Mary (Warner) and Bob Hooper, Lee Warner, and Bob Warner

Author's note: COR-TEN™ is a registered trademark of U.S. Steel

44 # Mended stockings
Louis (44) and Delia Trudell Miner

44 Mended stockings
Louis (44) and Delia Trudell Miner

Clifford Trudell with his grandfather's coffee cup, 2021. (Courtesy of Ann Trudell.)

[Author's note: Clifford Trudell passed away before this book could be completed, but it's important to retain the character of the interview as he experienced it.]

Clifford Trudell, Jr., is in his eighties now. He's one of the most senior descendants to provide information for this book. One of his most treasured possessions is the coffee cup and saucer that belonged

Trudell family ca. 1956. Back, Clifford, Irene, Joe, Rose, Lucille. Front, Edward, Delia, Joseph Trudell (brother of Louis), Loraine (in photo frame), Gladys. (Courtesy of Clifford Trudell.)

to his grandfather, Louis Trudell. A conversation with Clifford is like having coffee with someone you've known forever.

He said his Grandma Delia kept her feelings about the Barnes-Hecker tragedy deep within her. Perhaps she meditated on the tragic loss of her husband during her daily worship. Her devotion to her Roman Catholic faith was so strong that she walked to church and back every morning. She lived in the Salisbury neighborhood south of Ishpeming, about a mile and a half from her church, which was on south Lake Street. After breaking her ankle in the 1940s or 50s, she could no longer make that daily trek.

Clifford said Delia was very frugal and saved a bag full of worn stockings that she used for mending. She wore long flesh-colored cotton stockings, and he rarely saw her wear stockings that didn't have patches. She was generous with her baking skills, and often had sour cream cookies for her grandchildren. In the evening, she allowed herself the simple pleasure of a cup of Nestle cocoa as she sat in her rocking chair and let her mind wander to places that she never shared

Delia Trudell is pictured with Clifford Trudell (Sr.) and Lorraine behind her, and seated are Lucille, Irene, Gladys, Rose, and Edward, ca. 1926. (Courtesy of Clifford Trudell and the Marquette Regional History Center)

with her family. Clifford said he sometimes saw flickers of bitterness as she sat rocking and reflecting.

One person who has been missing from the Barnes-Hecker narrative is sixteen-year-old Joseph Trudell, the eldest of Louis and Delia Trudell's eight children. Joe had decided to be a hunting guide for out-of-town hunters, which he began when he was only 13. One of the hunters was so impressed that he hired Joe in 1926 to work in the town of Harbert, near St. Joseph in Michigan's lower Peninsula.

Joe left Ishpeming about a month before the Barnes-Hecker tragedy. He later owned a store in Harbert, where he provided summer jobs for his sisters Gladys and Lucille. Clifford Jr. also worked there for two summers during high school.

Another hidden figure in the Trudell history is Louis's brother Joseph Trudell, who became a grandfather figure for Clifford and his cousins. Joe served in the army in World War I, where Clifford thinks that Joe was exposed to toxic gases. He has a childhood memory of Joe

On Thursday, November 3, 2016, the 90th anniversary of the Barnes-Hecker tragedy, an ecumenical memorial service was held at Bethel Lutheran Church in Ishpeming. Clifford Trudell read from Revelation 21: 1-5, which includes the words, "…God himself will be with them; he will wipe away all the tears from their eyes." Clifford said it was one of the most meaningful experiences of his life. He passed away on September 14, 2021. It was a privilege to know him.

having such severe respiratory trouble that he could barely cross a street without being short of breath.

Joe's strength as a father figure was underscored when Louis and Delia's son Edward returned from serving in the Battle of the Bulge during World War II. Ed was so traumatized by the war that he wept in his Uncle Joe's arms.

Clifford Trudell (Sr.) was only 14, and the oldest son living at home when the Barnes-Hecker went down. He quit school to help support the family. He would allow himself only five cents for tobacco before giving his pay to his mother.

Louis and Delia Trudell's grandson Clifford Jr. worked for CCI at several different properties. After the Mather A closed, the buildings located on Malton Road were used for employee training in welding, cutting and the use of torches. While Clifford was working there, he was part of a repair crew for a large fan from the Pioneer Ore Improvement Plant. Part of the repair included use of a sealant that hardened like concrete. Because of a problem with the diagram for the fan, 104 cemented bolts had to be removed. That involved an air wrench and a 20-lb. sledge hammer. When the bolts yielded, they flew about 20 feet. One hit Clifford right between the eyes. He was knocked out, but okay.

Clifford said that dealing with dangerous situations is part of every job in mining. Unpredictable things can happen, as they did at Barnes-Hecker. Like Delia Trudell's stockings, people's hearts may carry patches that cannot be seen. But mending can happen, especially when the stitches are made from the threads of love that people share with one another.

–Clifford Trudell

45   Waterloo and a son's tribute
Louis and Delia Trudell

Edward Trudell addressed the audience with the monument behind him. (Photos in this chapter courtesy of Thom Skelding)

Edward Trudell was five years old when he lost his father Louis Trudell. Louis's brother Joseph became a father figure for his brother's seven children. The whole family called Joseph "Uncle." Ed's daughter Mary Trudell Borden said that as adults, she and her siblings realized how fortunate they and Grandma Delia Trudell were to have Uncle Joe in their lives and how blessed they were to have known him.

Uncle Joe taught Ed how to drive, and introduced the boy to the public library, where Ed expressed his amazement at seeing all the books. He vowed to read every single book in the library.

Ed reviews the names of the 51 lost men on the back of the monument.

Leo LaFond, left, hosted the dedication ceremony, with Ed at the podium.

Edward Trudell displays a photo of his father Louis at the monument dedication.

Perhaps this experience was the spark that led Ed to a lifelong career as an educator.

During World War II, Ed served in the Battle of the Bulge and as a medic in a military hospital in England. After the war, he took advantage of the G.I. Bill and enrolled at Northern State Normal School which is now Northern Michigan University. He then earned his teaching degree from Michigan State.

His first teaching job was 550 miles from home in Marine City, Michigan, where he spent his entire career as an educator. Within two years of being hired, he became principal of grades seven through twelve. Ed continued his own education and earned his master's degree in school administration from Wayne State University in Detroit.

Mary Trudell Borden said her father was loved by his students—he was strict, fair, and a lot of fun. He taught the cheerleaders the locomotive cheer from Ishpeming High School and led the girls in cheers at pep rallies. At a talent show, he arrived on stage with his hair combed like the Beatles and a guitar in his hands. As Ed mimed the words to "Waterloo," two students were backstage playing guitar and whispering the lyrics to him. When Ed sang the refrain for the second time, he brought down the house.

An appreciative audience gathered for the dedication of the Trudell monument at Cliffs Shaft Mine Museum.

In 2011, Edward Trudell fulfilled a lifelong dream to create a memorial to his father and the 51 lost men of Barnes-Hecker. Ed's nephew Leo Lafond served on the board of Cliffs Shaft Mine Museum, and he made arrangements to place the Trudell monument among the outdoor exhibits at the museum. On August 20, 2011, the audience was hushed as Ed gave an eloquent tribute to his father and the Barnes-Hecker men. The audience that day included two other sons of Barnes-Hecker families and many second-generation descendants and extended family members.

Mary said that a week before her dad died, the two of them reminisced about Ed's talent show performance and sang "Waterloo." "I still get choked up when I hear that song," she said.

When Ed died, scores of his former students came to pay their respects and tell stories about him. Ed had asked Mary to promise that Uncle Joe's American flag from World War I be placed into his casket and buried with him. Mary kept her promise.

In the ninetieth year since his father's passing, Edward Trudell died in the spring of 2016. He was the last known American member of the first generation of children of the lost men of Barnes-Hecker.

–Mary Trudell Borden and Julie Mosurak

46   **Other voices**
The conference of mining men

46   Other voices
The conference of mining men

The sweeping scope of iron mining in Michigan's Upper Peninsula encompassed thousands of men in mines on three iron ranges extending to the southern and western borders of the region. Serious incidents and loss of life had happened in many mines since the inception of iron mining in 1849.

In 1926, in Marquette County alone, there were 21 mines and five quarries that employed 3,931 men, with 2003 underground and 1393 on surface.

In the immediate aftermath of the Barnes-Hecker tragedy, CCI cast a wide net to gather a panel of experts to evaluate the disaster and lend their voices from an outside perspective. This chapter provides insight into their 1926 evaluations.

On Friday, November 5, 1926, CCI convened a conference of 24 mine leaders from Cleveland, Duluth, Ironwood, Iron Mountain, Ishpeming, Negaunee, Painesdale, Palmer, and the U.S. Bureau of Mines (now MSHA).

Their task: to visit the mine site, observe the recovery efforts, review maps of the mine, have an open discussion, and render an opinion on the feasibility of recovering the remains of the lost men and re-opening the mine. *The Daily Mining Journal* of November 6, 1926, stated that it was unprecedented for members of the media to be present at such a discussion.

The consensus opinion of the conference attendees was written and signed by the 12 men marked with an asterisk below. None of these 12 men were employed by CCI. The conclusions stated that CCI had operated the mine according to the accepted mining practices, and that the accident could not have been predicted or prevented. The document is included with this chapter, and was appended to the transcript of the February 1927 Coroner's Inquest into the deaths of the ten men whose remains were recovered.

- *FC Gregory, Asst. engineer, US Bureau of Mines, Duluth MN
- *OC Davidson, General superintendent, Oliver Iron Mining Co., Iron Mountain MI
- FE Keese, Oliver Iron Mining Co., Ishpeming MI
- Harry T. Hulst, Oliver Iron Mining Co., Ishpeming MI
- George Etsels, Oliver Mining Co., Iron Mountain MI
- *WH Schacht, General manager, Copper Range Co., Painesdale MI
- WJ Richards, Copper Range Co., Painesdale MI
- *Robert Walker, Consulting engineer, MA Hanna Co., Cleveland OH
- Capt. John Huhtala, MA Hanna Co., Palmer MI
- *William Kelly, US Bureau of Mines (now MSHA); formerly of Penn Iron Mining Co., Vulcan MI; past president American Institute of Mining and Metallurgical Engineers
- *Frank J Webb, General manager, Republic Iron & Steel Co., Duluth MN
- JE Nelson, Republic Iron & Steel Co., Negaunee MI
- Arthur Hansen, Republic Iron & Steel Co., Negaunee MI
- *AD Chisholm, Pickands Mather, Ironwood MI
- *WP Chinn, District manager, Pickands, Mather & Co., Duluth MN
- WA Rose, Pickands Mather & Co., Duluth MN
- Fred M. Prescot, Prescott Pump Co., Menominee MI
- Curry Prescot, Prescott Pump Co., Menominee MI
- *OM Schaus, Asst. range manager, Oglebay, Norton & Co., Ironwood MI
- *AE Richards, Superintendent, Ford Motor Co., Ishpeming MI
- Captain Holman, Ford Motor Co., Ishpeming MI
- JR Reigard, Ford Motor Co., Iron Mountain MI

- *Ralph S. Archibald, Michigan manager, CK Quinn Ore Co., Negaunee MI
- *Harry S. Peterson, Superintendent, Jones & Laughlin Ore Co., Ishpeming MI

The following men attended on behalf of CCI:

- S.R. Elliott – Assistant general Manager, CCI
- Capt Fred Ware – Negaunee Mine, (Inland Steel)
- Capt John Tregoning – Athens Mine, Negaunee
- WW Conibear – Safety engineer, CCI
- JH Rough – General underground superintendent, CCI
- Lucien E. Eaton – Superintendent, Republic Mine, CCI
- WW Graff – Superintendent, Gwinn District, CCI
- CJ Stakel – Superintendent, North Lake District, CCI

Leslie P. Barrett, Michigan state geologist and appraiser of mines, was traveling when he received word of the disaster. By the time he arrived in Ishpeming three or four days after the accident, his staff in Lansing had studied Barnes-Hecker records and maps, and he had been in telephone contact with CCI Assistant General Manager S. R. Elliott and other CCI officials.

Exhibit 1

At the invitation of the Cleveland-Cliffs Iron Company, a meeting of twenty four mining men representing various mining companies, operating in the Lake Superior Mining District, and the district representative of the U.S. Bureau of Mines, was held in Ishpeming, on November 5th, 1926, for the purpose of studying the conditions that led to the cave-in at the Barnes-Hecker Mine on November 3, and to consider methods that might be undertaken for the recovery of the bodies of the men who were buried.

After full discussion of the problem by all present a Committee was appointed to further study the question and to visit the mine and report on the conditions there. The Committee reports as follows:

The Barnes-Hecker mine was visited and the conditions on the surface inspected. It was found that that part of North Lake that could have affected the Barnes-Hecker Mine had been drained and all surface drainage led away from the mining area by extensive ditching. The maps of the underground workings were carefully examined and the history of the property and its operations from the beginning to the present time was given consideration. From this inspection and study it is our judgment that the mine was operated according to the best mining practice and that all care possible was taken by the mining company to safeguard their employees and that the accident was due to causes that could not have been foreseen.

It is the opinion of the committee that every man in the mine was buried and met death almost immediately and that rescue work was impossible as the mine workings were completely filled with water, sand and mud in a few minutes. The work undertaken by the Cleveland-Cliffs Iron Company to recover the bodies has been everything that could have been done and the methods adopted were the only means possible.

In the prosecution of this work it may be necessary to excavate and keep free from water the surface in and around the caved area. As the surface is very deep it will involve moving a large amount of earth and other material.

In conclusion it is the opinion of this Committee that the recovery work will require a considerable length of time.

William Kelly	Past president A.I.M. & M.E.
Frank C. Gregory	Ass't. Engr. U.S. Bureau of Mines
O.C. Davidson	Gen'l. Supt. Oliver Iron Mining Co.
W.H. Schacht	Gen'l Mgr. Copper Range Company
W.P. Chinn	Gen'l Mgr. Pickands, Mather & Co.
F.J. Webb	Gen'l. Mgr. Republic Iron & Steel Co.
O.M. Schaus	Ass't. Range Mgr., Oglebay, Norton & Co.
H.S. Walker	Consulting Engr., M.A. Hanna & Co.
H.S. Peterson	Supt. Jones & Laughlin Ore Company
Alvin Richards	Supt. Ford Motor Company
R.S. Archibald	Mich. Mgr. C.K. Quinn & Co.
A.D. ChisholmDis't.	Mgr. Pickands, Mather & Co.

47   The Coroner's Inquest
Cause and manner of death

On February 2, 1927, the Marquette County Coroner convened a jury tasked with hearing testimony about the *cause and manner of death* of the ten men whose bodies were recovered from the Barnes-Hecker. The jury was *not* tasked with rendering an opinion about the cause of the tragedy.

The transcript of the inquest is 66 pages, double-spaced and double-sided. It provides insight into the workings of the Barnes-Hecker and steps taken to assure the safety of the men. The witnesses had direct knowledge of seeing the men in the mine before the cave-in occurred, retrieval of the bodies, or knowledge of the conditions that existed before and after the cave-in.

Witnesses were questioned by County Prosecuting Attorney Clarence E. Lott, Coroner William Prin, and Thomas Clancey who represented CCI. Mr. Lott said that because Mr. Clancey was most familiar with the witnesses and conditions, he had asked Clancey to proceed with questions, and Mr. Lott, Mr. Prin, and members of the jury would ask whatever questions they desired.

The witnesses were:
- John Glandville (Glanville), night shift boss at Barnes-Hecker (See chapter 4: 5am)
- Allyvion Miners, underground track foreman (See chapter 8: A big wind in the shaft)
- Ed Hillman, pipe foreman (See chapter 7: It seemed as though the shaft was tore apart)
- Rutherford Wills, who climbed to safety (See chapter 6: Lone survivor)

- S.R. Elliott, CCI assistant general manager (See chapter 21: No warning)
- L.P. Barrett, state geologist (See chapter 24: Coring samples and a cork)
- William Conibear, CCI safety inspector (See chapter 11: I would have made my bed there)
- Charles Stakel, North Lake District superintendent (See chapter 10: Lanterns in the dark)
- William Kelly, U.S. Bureau of Mines (key points included in other chapters)
- F.E. Keese, General superintendent for the Oliver Mining Company in Ishpeming, who testified that he stood ready to bring his entire work force of 500 men to the Barnes-Hecker, but saw that there was no hope any of the men underground could be alive.

The jury verdict was that the deceased men had met their deaths in a cave-in of the Barnes-Hecker Mine, the cause of which was unknown.

The transcript is signed by Sybil I. Gingrass, clerk of the court, who recorded the testimony.

It was an open hearing, and on February 3, 1927, *The Daily Mining Journal* ran a lengthy article with the headline: "Mine Inquest Jury Returns Open Verdict" that accurately summarizes the inquest, but contributed to the decades-long misunderstanding of its purpose:

> ISHPEMING Mich., Feb 3—The inquest *into the Barnes-Hecker disaster* [emphasis added] of November 3, 1926, when the lives of 51 miners were lost in less than 15 minutes by the rush of sand and water from a surface cave-in at the property, was held at City Hall and the jury returned the following verdict: "That the deceased met their death in the Barnes-Hecker Mine by a cave-in, the cause of which is unknown."

The "deceased" in context means the ten men whose remains were recovered, and again, the purpose of the inquest was to present evidence and obtain a jury verdict on the *cause and manner of death,* not the cause of the accident.

A copy of the inquest was handed down to me in the 1990s, and I first read it at that time. Since then, my copy has become dog-eared, flagged, underlined, battered, and marked with a lot of penciled notes.

No single document about the Barnes-Hecker tragedy could possibly tell the whole story. The focus for this book is the experiences of the people. The words of the witnesses convey their unspoken sense of disbelief, dismay, frustration and angst in the face of "something peculiar," "something entirely out of the ordinary," something for which they had no explanation that had defied their ability to predict or prevent the worst mining disaster in Michigan history.

Because no other chapter provides insight into William Kelly, his obituary from "Proceedings of the Lake Superior Mining Institute," 1939, p. 38, is included here:

> William Kelly, a charter member of the Institute, died October 9, 1937, at the age of 83, at his home in Vulcan, Michigan.
>
> He was born in the East and attended Columbia University, going to Iron Mountain, Michigan, after graduation. He was one of the few technical men in the district when he engaged in mining work, stepped up rapidly in his chosen profession and became head of the mining interests of the Penn Iron Mining Company on the Menominee Range. The Company was owned by the Cambria Steel Company until 1923, when the Bethlehem Steel Corporation took it over and from that date was operated by Pickands, Mather, and Co.
>
> In 1924, Mr. Kelly served as president of the American Institute of Mining and Metallurgical Engineers. This was a year after his retirement from active work. In 1898 he was honored with the presidency of the Lake Superior Mining Institute. For many years he was a member of the Board of Control of Michigan College of Mining and Technology [Michigan Tech University] and chairman of the board for the greater part of that time.
>
> His wife survived him. Burial took place in Vulcan.

48 The Inquest jury
Coroner's inquest jury

In our quest to seek out the people involved in the Barnes-Hecker tragedy and its aftermath, we wanted to learn more about the men who served on the coroner's inquest jury. On November 8, 1926, Coroner William Blewett released the names of the six men who would serve on that jury. They were Erick Bergdahl, David Creagle (Crago), James Murphy, William Rashleigh, John Skewes, and John Webster.

It's essential to understand that the purpose of the inquest was solely for the jury to render a verdict on the *cause and manner of death* of the ten men whose remains were recovered. It was not about the cause of the Barnes-Hecker tragedy. Prior to the advent of forensic pathology, coroners could convene inquest juries to ponder the circumstances surrounding deaths that happened in uncertain circumstances.

At the conclusion of the inquest, *The Daily Mining Journal* ran the headline, "Mine Inquest Jury Returns Open Verdict." The article correctly quotes the verdict "that the deceased met their death in the Barnes-Hecker Mine cave-in, the cause of which is unknown," but the lengthy article leaves the impression that the inquest was about the cause of the tragedy, which it was not.

The ten men recovered from the Barnes-Hecker were Henry Haapala, Jack Hanna, William Huot, Thomas Kirby Jr., Thomas Kirby Sr, Nels Hill, Marquette County Mine Inspector William Hill, Joseph Mankee, William F. Tippett and Arvid Wepsala.

This chapter provides biographical information about the members of the jury.

Erick Bergdahl (1849-1946). Bergdahl had immigrated from Sweden in 1881. "A stonecutter and mason, he followed that trade here, and many of the stone buildings still standing in the city are his work." Mr. Bergdahl was 97 at the time of his passing and was survived by a son, three daughters, 18 grandchildren and 12 great-grandchildren.

David Creagle (Crago) (1877-1933). After an exhaustive search by the Longyear Library at the Marquette Regional History Center, no information with the spelling Creagle could be found, but they did locate an obituary with the spelling **Crago**, dated October 30, 1933. He had lived in Ishpeming for about 30 years. As a young man, Mr. Crago was known as an expert mine driller. Coming to the Copper Country from England, he worked for a number of years at the Quincy Mine. During contests held at Houghton, Crago won two first-place awards for drilling in one day. He later moved to Ishpeming, was employed at the Ropes Gold Mine and helped to sink the shaft at the Cambria Mine in Negaunee. Crago was credited with saving the life of his partner in an Arizona mine when a large electric drill fell on the other man. Crago lifted the drill off of his partner, but injured himself in the process. He became paralyzed on one side of his body. He was hospitalized in Arizona for ten months before returning to Ishpeming. According to the obituary, he lived for several years at the Billing Hotel. He was unmarried and had a brother John in Hancock. (Author's note: Mr. Creagle/Crago's paralysis might have affected his speech, and this, in turn, could have led to incorrect spelling of his name.)

James Murphy (1860-1931). Murphy had immigrated from County Tipperary, Ireland, in 1880 and resided in Ishpeming for more than 50 years. He was 71 when he died of a stroke in 1931. His obituary hailed him as pioneer in mining and one of Ishpeming's best-known residents.

He had worked as a pump man for the Oliver Mining Company for 46 years. He was predeceased by his wife Margaret, and survived by a son, two daughters, seven grandchildren and a sister.

William Rashleigh (1882-1957). A native of Sweden, Rashleigh's 1918 World War I draft card lists his occupation as a mechanic at the Lake Angeline Mine, and his obituary states his occupation as an electrician. He died the age of 75, and for the last ten years of his life, he worked as a bartender at Beanie's Bar, which was operated by his brother Richard.

[Author's note: There were two men named John Skewes who could have served on the inquest jury. With appreciation to the Longyear Library at the Marquette Regional History Center, both men are presented in this chapter.]

John W. Skewes (1880-1961). Born in Marazion, Cornwall, England, 80-year-old Skewes had arrived in the United States in 1892 and Ishpeming in 1894. He had been employed by the Oliver Mining Company and Cleveland Cliffs Iron Company until he retired in 1947. The 1930 census listed his occupation as a teamster in an iron mine. He was survived by a daughter, four grandchildren, one great-grandchild, one sister and one brother.

John Richard Skewes (1885-1969). John Richard Skewes was born in 1885 in Truro, Cornwall, England, and worked for the Oliver Mining Company until 1947. A 60-year resident of Palmer, Michigan, he was survived by several cousins in England. He died in 1969.

John Webster (1870-1952). A native of Nova Scotia, Webster immigrated when he was eight years old. His obituary states that he had been an engineer for many years at Ishpeming High School. In ill health for a number of years before he died, he was survived by three sisters.

49   Sub-level caving
by the top slicing method

In the 1927 inquest, the question was asked about the mining method in use at Barnes-Hecker. S.R. Elliott replied, "the caving system is the safest system that has ever been devised."

In every aspect of the Barnes-Hecker story, context matters. S.R. Elliott pointed out that the caving system would ultimately break the surface, but in a controlled and predictable manner. Over many years on the Marquette Iron Range, this knowledge led to proactive movement of houses out of areas where caving was predicted.

In the 1925 proceedings of the Lake Superior Mining Institute, George Newett, geologist and publisher of the *Iron Ore* newspaper, provided an overview of mines on the Marquette Iron Range. His comprehensive paper includes mining methods in use at CCI mines including the years when shafts were sunk. His report also includes data on the drifting record set at the Barnes-Hecker in March of 1920.

- Barnes-Hecker (1917): *sub-level* caving by the top slicing method [Ishpeming, MI.]
- Morris (1909): sub-level caving by the top slicing method. Shaft was sunk in swamp by the Foundation Company using the Caisson method to make a seal at ledge. [Ishpeming, MI.]
- Lloyd (1909): sub-level caving by the top slicing method; connected to Morris [Ishpeming, MI.]
- Holmes (1916) sub-level caving by the top slicing method; connected with Oliver Mining Co. Section 16 Mine for ventilation and safety. [Ishpeming, MI.]
- Athens (1913-1917): sub-level caving by the top slicing method [Negaunee, MI.]

- Negaunee shaft 3 (1908): sub-level caving by the top slicing method [Negaunee, MI.]
- Maas (1902-1906): sub-level caving by the top slicing method. Most difficult drop shafts ever completed on the Marquette Range. [Negaunee, MI.]
- Stephenson (1905): sub-level caving by the top slicing method. [Gwinn, MI.]
- Princeton (operated since 1872, acquired by CCI 1905): sub-level caving by the top slicing method [Gwinn, MI.]
- Gwinn (1907-08): sub-level caving by the top slicing method. First shaft to be sunk using the Caisson method with a seal at ledge to exclude surface water. [Gwinn, MI.]
- Austin (1903): sub-level caving by the top slicing method [Gwinn, MI.]

Newett's paper also included the names of key personnel at CCI:

> The representatives of the company on this range are: M. M. Duncan, vice president and general manager; S. R. Elliott, assistant general manager; George R. Jackson, superintendent Negaunee group of mines; Lucien Eaton, superintendent Ishpeming group; Charles Stakel, superintendent North Lake group; W. W. Graff, superintendent Gwinn group; W. R. Myers, superintendent Republic and Iron River properties. James H. Rough, head mining captain. The general offices for the range are located in Ishpeming. James E. Jopling is chief mining engineer; Edwin L. Derby, Jr., chief geologist; O. D. McClure, chief mechanical engineer; A. J. Yungbluth, purchasing agent; C. J. Shaddick, assistant auditor; William Conibear, safety engineer; W. H. Moulton, head of welfare department; Fred Baker, chief chemist; H. R. Harris, of Marquette, is the vice president and general manager of the railway department; Austin Farrell, of Marquette, the manager of the furnace department, and F. W. Hyde, of Marquette, superintendent of lumber operations. The following mining captains are employed by the company on the Marquette Range: Alfred Collick, James Stephens, William Tamblyn, John Olds, William Tippett, William Nault, Peter Pascoe, Joseph

Thomas, Fred Ware, John Tregonning. William Jory, August Fagerberg and Alfred Bone.

A DRIFTING RECORD. Records in shaft sinking and drifting are subjects that are full of interest and create much discussion. I have secured the following facts from the company's books, of a big job of drifting at the Barnes-Hecker mine, of the Cleveland-Cliffs Iron Company, done in March, 1920. John M. Bush was in charge of the property for the company, and Captain William Tippett was the mining captain. The drift was 9x10 feet, in compact slate, with much water. A mechanical loader was used, and the ventilation was excellent:

Number of feet drifted:	539
Number of days worked:	27
Number of feet per day:	9.33
Number miners in drift:	4+
Number continuous hours worked:	552
Number of feet per hour:	0.987
Number of hours worked by miners:	2,280
Number of hours worked by trammers:	2,414
Number of miners hours per foot:	4.23
Number of trammer hours per foot:	4.48
Tons of rock per foot of drift:	9.70

In 1936, in a paper published in *Proceedings of the Lake Superior Mining Institute*, William Conibear wrote:

> Nearly all the iron ore mined in Michigan is obtained underground, and top-slicing and sub-level caving are the principal methods of extraction. These mining methods are particularly favorable for guarding against accidents by falls from roof or *back*.

Contemporary sublevel caving is the topic of two chapters in the 2001 mining textbook *Underground Mining Methods, Engineering Fundamentals and International Case Studies* by William A. Hustrulid, and published by the Society for Mining, Metallurgy, and Exploration, Inc. (SME) in Littleton, Colorado.

The chapter titles are, "Sublevel caving: A fresh look at this bulk mining method," by C.H. Page and G. Bull, pp. 385-393; and "Underground Iron Ore Mining at LKAB, Sweden," by C. Quinteiro, M. Quinteiro, and O. Hedstrom, pp. 361-384.

Depending on access rights, these chapters can be viewed online. Interested readers are encouraged to search by the chapter titles.

Drifting Record

It is a virtual certainty that this widely recognized Barnes-Hecker photo was taken to commemorate the milestone drifting record set in March of 1920. If this is true, then the equipment is the mechanical loader used to achieve the 539-foot record. At left is pump man Peter Mongiat; next to the loader is Olaf Trondson; crouched in front is Mine Captain Bill Tippett; the man on the right is likely to be Allyvion Miners, who was both shift boss and drifting boss. The seven other men were miners and trammers.

Articles in the October 11, 1919, and May 22, 1920, issues of the *Iron Ore* newspaper document the use of an Armstrong Shuveloader to achieve the 1920 drifting record. This photo published in a U.S. Geological Survey report shows the loader in action ca. 1922 in the Congo Phosphorous Mine in Idaho. (Photo courtesy of Idaho State Historical Archives)

See also: Chapter 37 No longer anonymous

50   History of the mine site

This is the tripod used during the initial phase of shaft-sinking in 1917. Four men are dwarfed next to the left leg of tripod. (Courtesy of Marquette Regional History Center.)

One of the great joys of working on this book has been serendipitous discoveries that have come to light. One of those discoveries was a small envelope that had fallen to the bottom of a box of documents at the Longyear Library at the Marquette Regional History Center. It's exciting to present the photos from that envelope, as well as additional views from the Marquette Regional History Center collection, and photos from Greg Montgomery who now owns the private collection of photos from Allan Koski.

Top is a rare photo of the head frame after the mine site had been cleared of timber in the fall of 1917. The headframe from the closed Chase Mine had been stored at the Morris Mine, and CCI hired the Worden Allen Company to install the structure at Barnes-Hecker. If new steel had been available during World War 1 in 1917, it would have cost more than a million dollars. Note the large concrete anchor in the foreground. There were pairs of these anchors that secured the cables on each side of the single-track steel trestle supports in the photo below it. (Top photo Courtesy of the Greg Montgomery/Allan Koski collection; bottom photo courtesy Marquette Regional History Center has been previously published.)

The headframe ca. 1918 facing west. Notice there is not waste rock under the trestle that goes south (left) of the headframe. The photo appears to predate the view bottom ca. 1918-19. The timbers stacked around the headframe were used underground. (Top photo courtesy Greg Montgomery/Allan Koski collection; bottom photo courtesy of Marquette Regional History Center)

Facing south, the foreground shows preparation for a temporary engine house used while the shaft was being sunk. (Courtesy of Marquette Regional History Center)

Facing west, 90 degrees from the previous photo, this view shows the boiler and dry building, with the excavation for the engine house in the foreground. In 1918 The *Iron Ore* stated that the dry, where men changed into mining clothes, boasted steel lockers and modern showers with concrete floors. The building is called a dry because it is equipped with hooks and baskets suspended by chains from the ceiling. Each man was assigned one of the hooks for wet mining clothes that were hoisted up near the high ceiling to facilitate drying. (Photo courtesy of Marquette Regional History Center)

This is the base of the headframe facing north. The building next to the shaft is the shop. (Used with permission of Marquette Regional History Center)

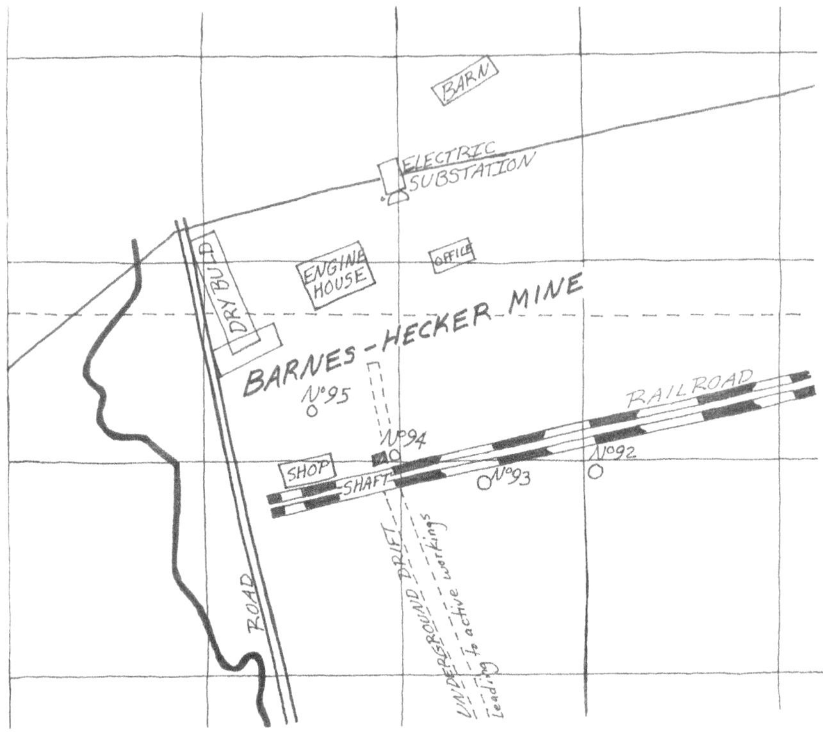

This illustration shows the positions of the Barnes-Hecker structures in 1926, based on the original map in the collection of the Central Upper Peninsula and NMU Archives. Drawn by J. Walitalo.

See also the Appendix for a map.

51 | The Barnes before Barnes-Hecker

Seven years before the advent of work on the Barnes-Hecker, there was a small, short-lived Barnes Mine that existed a few hundred feet to the northwest. The known dates of its existence are 1909–1912, and possibly a little longer. The captain was Alfred Rowe, whose position is documented in the 1912 financial report. The North Lake *District* Superintendent was John Bush.

Both the Barnes and Barnes-Hecker mines were on land leased from the Barnes Land Company. The shaft of the Barnes Mine was only 350 feet deep, while the Barnes-Hecker shaft was 1,050 feet deep. The Barnes Mine apparently had no reportable ore production.

On November 10, 1910, an unnamed miner was hit by a loose timber that fell down the shaft. He died the following day.

The records for the Barnes Mine may shed light on some missing pieces of the Barnes-Hecker story.

If CCI routinely had crew photos done on the last day of mine operations as it did for the 1979 closure of the Mather B, then this photo would have been taken at the Barnes ca. 1912, and predated the Barnes-Hecker disaster by 14 years.

In 1910, John Bush was succeeded by W. W. Graff as superintendent of the North Lake District. In the meantime, Bush moved to Ashland and then the Iron River District before returning to North Lake, probably in 1916 when Graff moved to the Gwinn District. In 1925, Charles Stakel became superintendent at the North Lake District.

The Barnes Mine story continues in the next chapter.

                                                  Mary V. Tippett

In this photo at far left is Emil Maki, fourth from left William Tippett, fifth from left is Tippett's son Will. Both Emil Maki and William Tippett perished in the Barnes-Hecker tragedy. (Photo courtesy of Tippett and Maki descendants.)

52   The Barnes-Hecker Location
A tale of one neighborhood in two phases?

The homes in the Barnes-Hecker Location were ready for occupancy. In 1918. The mine captain's house stands out on the left side and a duplex is visible on the right. (Courtesy of anonymous)

Throughout the research for this book, there have been inconsistent descriptions of the Barnes-Hecker Location, which was a small neighborhood of company housing located about 1800 feet north of the mine.

At the time of my original research of CCI records, the digitized records for the 1909–1912 Barnes Mine and its Barnes Location weren't available in the online collection of documents in the Central Upper Peninsula and Northern Michigan University Archives.

This 1918 photo shows the outdoor pump that was believed to be the sole source of water for the entire neighborhood. Note the tall trees that were left in place to enhance the landscape. (Courtesy of anonymous)

It appears that the Barnes-Hecker (housing) Location may have been built on the same site as the earlier Barnes Location, which had five duplexes, the captain's house and a boarding house. The Barnes-Hecker Location had ten duplexes, the captain's house, and a one-room school. The site was actually closer to the old Barnes Mine than to the Barnes-Hecker.

The infrastructure already in place from the older Barnes Location included a road, footprints for five duplexes, a captain's house, and four drive wells that produced an excellent water supply. The description in the 1909 records praised the graded road and attractive yards boasting grass, flower beds, and fences like the ones in the North Lake Location. The Barnes Location was said to be a source of pride for North Lake District Superintendent John Bush.

CCI placed so much emphasis on having well-kept housing that it gave annual awards for the best kept flower gardens, flower boxes and vegetable gardens.

Another view of the neighborhood shows the tall trees that shelter the homes. (Courtesy of anonymous)

When the Barnes-Hecker Mine was being opened, the 1918 Ishpeming *Iron Ore* ran a glowing write-up of the Barnes-Hecker Location, stating that the residents would have guidance from a landscape artist in beautifying their yards; however, it wasn't until the fall of 1926 that stumps were blasted out of the road, the yards had grass and flowers, and the neighborhood had an attractive appearance.

The article also mentions that the Barnes-Hecker Location was *east* of the Barnes-Hecker, but it was actually *north* of Barnes-Hecker and *east* of the old Barnes Mine site.

The lore connected with Barnes-Hecker was that there was a single outdoor hand pump to provide water for the entire neighborhood, but that belief ran contrary to all the records about CCI concern for the welfare of the men and their families. It would have been impossible for one hand pump to serve the needs of 21 households. The 1920 Census shows there were 112 people living in the Barnes-Hecker Location.

A close-up of the school that was moved to the site from the former Dexter Mine Location in 1922, and the captain's house.

"Barnes-Hecker Location 1918-1927" details from large illustrations by the late Roger Junak, used with permission of the Michigan Iron Industry Museum.

The second close-up is a detail of family life in the Barnes-Hecker Location.

The more plausible scenario was recorded by Angelo DiBernardo, grandson of shift boss Sam Phillippi and his wife Mary in the chapter "Man's Impermanence." He wrote that there was one hand pump for every two households. It's also possible that if five duplexes were built on the footprints of the old Barnes duplexes, they could have been supplied by the four drive wells touted in the 1909 CCI records.

John Bush, who was superintendent of the North Lake District when both the Barnes and Barnes-Hecker Locations were built, would have known about the existing wells, which might have supplied water to some of the duplexes.

When flooding occurred in the Barnes-Hecker in 1920, a large number of men left. In 1921-22, while North Lake was being drained and a drainage ditch was dug, the payroll dropped to as few as 15 men, all listed as pumpmen. Records show that because of a drop in demand for iron ore, many CCI mines closed from June of 1921 until June of 1922, and there were layoffs

After the water problems were resolved in 1922, the payroll rebounded and the mine began producing ore on November 1, 1922.

In July of 1924, when the Salisbury Mine closed, the Barnes-Hecker hired enough men for a second shift. It's unknown how many residences were occupied in the interim.

In 1925, water issues in the Location came to the attention of newly appointed North Lake District Superintendent Charles Stakel. The men said their wives were frustrated at the inadequate water for laundry days on Mondays, and Captain Tippett said his house had a kitchen sink and bathtub, but no running water. Stakel obtained permission to have running water plumbed into the homes.

At the Michigan Iron Industry Museum, there are two illustrations commissioned by Lyle LaBeau Tippett, who lived in the Barnes-Hecker Location when she was five years old. That would have been 1923. The illustrations by Roger Junak show an idyllic place, not one with stumps in the middle of ungraded roads.

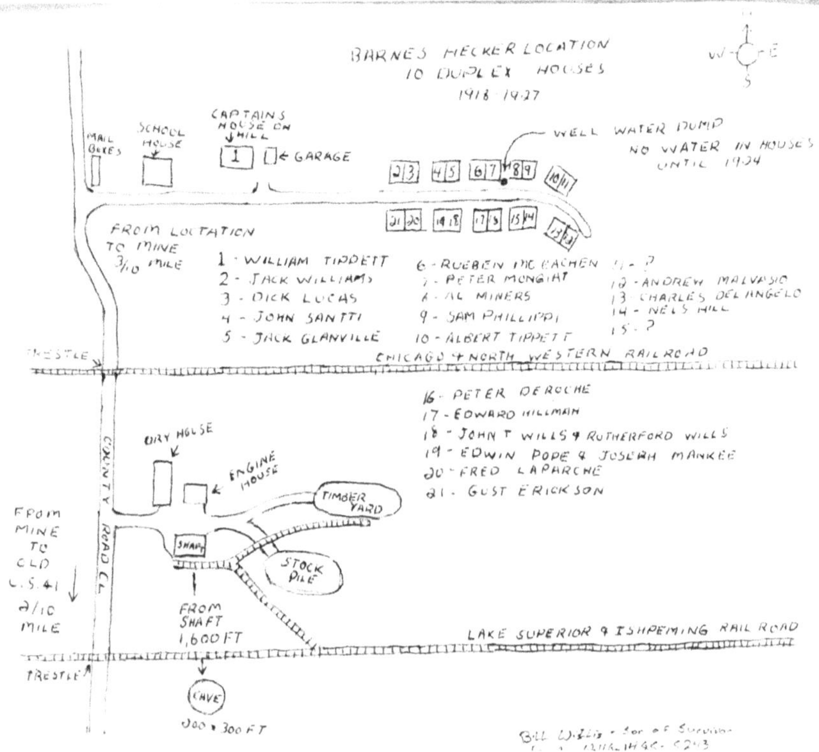

Map drawn by William Wills (Rutherford and Bruna's son). Although Wills's drawing has two names missing, other records show that Tom Drew would have lived in number 11, and Gust Erickson would have lived in number 21. The cave-in was 1,600 feet southeast (not south as shown).

Could it be that the Barnes-Hecker Location was simply a neighborhood where a portion was neatly finished while the other waited until 1926 to be beautified... a neighborhood in two phases? Regardless of the answer, the people who lived and visited there had many cherished memories of happier times.

Here are the names of the men living in the Barnes-Hecker Location at the time of the tragedy. The men who perished are indicated with *: electrician Charles Dellangelo, pumpman Peter DeRoche*, miner Tom Drew*, brakeman Gust Erickson, shift boss John Glanville, motorman Nels Hill*, motorman Fred (listed as Edlore on payroll) LaParche, trammer Joseph LeBoeuf, hoist operator Richard Lucas, cage rider Andrew Malvasio, miner Rueben McEachen, track foreman Allyvion Miners, pumpman Peter Mongiat*, shift boss Sam Phillippi*, miner Edwin Pope and stepson skip tender Joseph Mankee,* miner John Santti*, pump helper Albert Tippett, Mine Captain Bill Tippett*, pumpman R.J. Williams, dry man John Wills and motorman Rutherford Wills.

Only eight of these were listed in the 1920 census: Dellangelo, Miners, Mongiat, Williams, Wm. Tippett, Laparche, Hillman, and Lucas.

It's impossible to determine how many people lived in the Barnes-Hecker Location at any given time, except that at the time of the tragedy 33 were school-aged children.

53 Man's impermanence
Sam (44) and Mary Phillippi (Sante Filippi)
Shift boss

This photo shows one of the duplexes in the Barnes-Hecker Location in the fall of 1918 when the homes were newly completed. (Photo courtesy of anonymous.)

On the morning of November 3, 1926, Shift Boss Sam Phillippi was underground making inspection rounds with Mine Captain Bill Tippett and Marquette County Mine Inspector William E. Hill. When tragedy struck, Sam was caught underground. His body was never recovered.

Two months earlier, Sam had given his daughter Bruna in marriage to Rutherford Wills, who was a half-brother to mine captain Bill Tippett. Wills was the only man to escape the flooding mine alive.

Sam and Mary Phillippi lived in the Barnes-Hecker Location with their son, John, who was six years old. Sam's accident report said there were four surviving children. In addition to Bruna, they had a married daughter, Judith (Julie) DiBernardo, who lived in Pennsylvania. Judith and Oreste DiBernardo's son Angelo wrote a manuscript in 1945-46, and a copy was handed down through the descendants of Walter Tippett.

What follows in this chapter is excepted directly from Angelo DiBernardo's manuscript. It provides insight through the eyes of his grandmother, Mary Phillippi, about life in the Barnes-Hecker Location prior to the 1926 disaster, as well as a brief peek into her life afterward.

> Grandma Filippi had another mass said in memory of her dead husband today. That makes 19 annual masses since he died in the Barnes-Hecker cave-in up north. I was two years old when the mine went down so I never knew Grandpa Filippi—Sam the men at the mine used to call him. He was a shift boss and shouldn't have been in the mine so near lunch time when the accident hit; but since Mr. Hill, the county inspector, was making a routine check of the mine that day, he was down there too and trapped with the rest of them.
>
> My mother used to tell me about how we traveled from our home here in Pennsylvania up to the northern peninsula of Michigan at the time of the funeral – only there was no funeral for Grandpa, his was one of the 44 [sic] bodies that were never recovered. [41 were not recovered]
>
> Grandpa came from Italy and went up to work in the Marquette iron range so that he could make stake money and send for his family. (That's why I'm here in the states now

instead of hell knows where.) I owe a lot to Sam Filippi—besides my mother.

After working in a few of the different mines in the area, Sam Filippi went to work for Cleveland-Cliffs Iron Co. [in 1920] at their new Barnes-Hecker mine out in the woods about seven miles from Ishpeming, the iron city that sits some 15 miles away from the Lake Superior ore port of Marquette.

The family lived in the Barnes-Hecker location. A location is a small group of houses that the company builds near the mine so the men who work there won't have to stay in town, which in this case was some seven miles away. If you owned a car or owned your own home in town it was different; about 20 families lived in the location and all the men worked at the mine. There were some from Ishpeming and surrounding farms who worked in the mine too. They had to travel every day.

Each family lived in one half of a double, gray frame, slant roofed house, except for the Mine Captain, William Tippett, (also killed in the cave-in) who lived in a single house just beyond the mailboxes as you swung off the road into the location clearing. The double houses were plenty roomy. There was a living room, dining room, and kitchen downstairs; two bedrooms and an attic above stairs; a cellar dug out under the kitchen; and two big porches clean across the front and back of the house. The houses stood in two rows of five looking at each other across the road from behind the rows of pines that fronted them. This clearing in the woods also had a green school building.

Living was good on the location. Everyone got along, despite the fact that there were Finns, Italians, Frenchmen, and Englishmen all mixed together. The men worked the mine together and spent their leisure time playing together. The women traded visits and recipes—except those special recipes for meat pasties. They took pride in their pasties, those individual meat pies that are still popular all over the north country and anywhere else north country people have spread. The men even had their women carry pasties to the mine so that they could eat them warm for lunch; the lunch pails had special pasty compartments.

While the men were away at the mine, the women buzzed around their kitchens. The kitchen is always the coziest and most used room in a working man's house; just as a rich man uses his parlor or living room. For a while the kitchens were awkward because all the water had to be carried from pumps outside, one pump to every two houses; but after a while, pipes were run back and water ran right into the kitchen sinks. All the cooking was done over wood fires [wood-fired kitchen stoves] and this made wood-chopping a chore that always kept coming up. The wood was stored in a back yard shed that also served for a poultry house and in some cases a barn, for every family kept chickens, and my grandma and the Hills both kept a cow—that meant milking and feeding every morning in addition to the other chores. [The family of Nels Hill, who also perished in the tragedy, lived across the street in the Barnes-Hecker Location.]

The square-built houses were easily heated through the long Michigan winters by an anthracite burning pot-bellied stove that stood in the living room. Every one gathered around it on winter evenings, dreaming mining-town dreams in the flames that flashed from behind the isinglass windows in the sides.

Grandpa got his deer every year, and he always got it in the same place, a run that topped a small rise about fifteen minutes from the location. He would sit on a log and wait for an obliging buck to bound into his sights and grace his table with venison. Shortest time out and back for a buck was 35 minutes one year.

Pays at the mine ranged from $3.50 to $6.00 a day with correspondingly low living expenses. House rent and all the electricity you burned in one month cost only $9.00; a good bit of food came from the gardens, the woods, or livestock; store-bought food, delivered daily by two stores in Ishpeming, was not expensive; and clothing costs were low.

Most of the families could afford to send their children on to high school after they finished the eighth grade at the one-room schoolhouse. Pupils, ranging from first grade to the eighth, recited in one room under one teacher who was just out of the Normal School [Northern State Normal School, now Northern Michigan University]. Some of the boys, however, left school for the mine before they were all the way through.

The mine was always in the background of everyone's mind. Men and women lived in unspoken dread of accidents. The surface was calm, but there was always an undercurrent ready for an emergency.

News of trouble needs no carrier to spread—everyone seems magically attuned to it. My Aunt Bruna (she had been Wilfred Wills's wife for about two months) was on her way to the mine with lunch for Wilfred and Grandpa Filippi when she heard the shaft's whistle. A neighbor who had driven into Ishpeming that day passed her on the road and told her the mine had gone in. Bruna was frantic. When she got to the mine, she wasn't allowed to see him because the men were working on him in the dry house; an 800-foot climb in ten minutes is not relaxing. Women started to come out from the location. At first there seemed some hope that the men might be in an underground pocket and safe; but this soon died and grief set in completely. The red-eyed women and children just milled around, stunned into dumbness. They couldn't even sympathize with one another.

Nothing seemed real. The women couldn't get used to the idea that their men weren't coming back from the mine this time. The location was quiet. Sobbing children couldn't understand what had happened.

My mother and I got there two days after the accident. There was nothing we could have done. After a month or so, my grandmother came back down to Pennsylvania with us and stayed until the following June. When she went back up north she lived at the location for several weeks while the company repaired the house she was to move into in North Lake, some five miles away. There was only one other family living at the location.

Last summer we drove to out to the old shaft, the location, and the cave-in lake. No one has been to the lake by car for quite a while. The twin ruts running through the woods were fading. The lake was still. I didn't even see a bird. I threw a piece of gravel out into the water and while I watched the water ruffle up, I mumbled a prayer for the men who were under me somewhere.

The shaft was equally deserted. All that remained was the rusting superstructure. The buildings are gone; and the tracks ripped up. Rotting planks cover the water-filled shaft hole. I dropped a stone between two of them; it clanged on a pipe, hit the stone shaft side, and made an echo in the water.

Long grass covers the location clearing. A crumbling cellar foundation is the only sign that houses once stood there. Every building has been torn down. I couldn't even find a wood splinter left behind. The "prettiest little location you would ever want to see" is just a space of grass where cows sometimes graze. Man's impermanence becomes overbearingly obvious at a place where nothing remains of life a mere twenty years later... Philosophizing about the past doesn't affect the present at all.

The miners don't talk about the Barnes-Hecker very much; in fact, they shun all mention of the accident. Oh, they'll answer direct questions; they're too polite by nature to refuse. But their attitude is, 'We know that things like that can happen. Why make them worse by talking about them?' The men still go down into the mines day after day—and their women still wait for them. Mining is easier now, a lot safer, too; but prayers are still said at night and the shadow of Barnes-Hecker still lingers.

– Angelo DiBernardo

54   From shadow to sunlight

The Barnes-Hecker Mine site has been on private property ever since the time when the mine was operational. At the time of the 90th anniversary Barnes-Hecker Remembrance, the site was forested and silent, almost eerie in late afternoon shadows. In recent years, the timber was harvested and an interesting thing happened. Birds began to sing, a ruffed grouse ran across the trail, and a fleeting flash of a deer's tail disappeared into the new growth of poplar trees ruffled by the wind. It felt as though the souls of the lost men had been set free. Seasons change, sunlight chases away the shadows, and we are constantly reminded of the impermanence of humankind.

I hope this book is worthy of the lives of the people caught up in the maelstrom of the Barnes-Hecker tragedy.

This work is the result of marvelous collaboration with more than 50 people from the across the country including California, Colorado, Florida, Iowa, Kentucky, Maryland, Massachusetts, Michigan, Minnesota, Missouri, New York, North Dakota, Oregon, Pennsylvania, Texas, and Wisconsin, and people in Belgium, the British Isles, Finland, and Germany.

Some serendipitous connections have felt transcendent, and surely there have been some guardian angels watching over all of us. Breezes ruffle the surface of the lake that formed in the caved ground of the Barnes-Hecker, and I like to think the sunlight shimmering off the water is like a sparkling echo of a chiming bell that rang through the forest in 2002; an echo that reminds us there are no tears in heaven.

To my collaborators: bless your hearts for all the time you've spent digging for photos, rediscovering family histories, preparing note cards for our conversations, reviewing and refining drafts of our chapters,

PETER WHITE MEMORIAL AWARD
Presented to the
BARNES - HECKER
REMEMBRANCE COMMITTEE
For the preservation and enhancement
of history in Marquette County
PRESENTED BY THE
MARQUETTE REGIONAL HISTORY CENTER
-2016-

State History Award
Historical Society of Michigan
Barnes-Hecker Mine
Remembrance Committee
Special Programs/Events
September 21, 2019

and for lively and engaging conversations that have brought some of us to tears.

In 2016, the members of the Barnes-Hecker Remembrance Committee navigated together through an overwhelming response from the public, the media, and people who offered us more than we could have asked for. My hat is off to former Ely Township Superintendent Carl Hosang, United Steelworkers Locals 4950 and 4974 Representative Greg Montgomery, Cliffs Shaft Mine Museum Archivists Irene Ilmonen and Suzie Saari, Past President of the Board of the Marquette Regional History Center James Paquette, Cliffs Shaft Mine Museum Historian Gary Vidlund, Negaunee Historical Society President Virginia Paulson, Past President of the Ishpeming Historical Society Susan Phare Boback, and Michigan Iron Industry Museum Historians Troy Henderson and Barry James.

Our work has been honored with the 2016 Peter White Award from the Marquette County Historical Society and the 2018 Award for Best Special Project or Program from the Historical Society of Michigan.

These awards honor all the people who were swept into the inundation and collapse of the Barnes-Hecker Mine: people who survived; people who suffered; people who grew up without their dads, grandfathers, uncles, brothers, sons, husbands, and loved ones. One of the most poignant and universally felt legacies is that we couldn't know these men personally, and it is my fervent hope that through this book, we can celebrate their lives in the sunlight, with a sense of appreciation and gratitude.

With a grateful heart, I am sincerely,
Mary V. Tippett

Chairperson, Barnes-Hecker Remembrance Project
Granddaughter of Walter and Marian Tippett (Hill)
Daughter of Dewey and Ellen Tippett
Proud Mom and Grammie to James, Laura, and Reid

Glossary

Back: Ceiling or roof of an underground mine

Bulkhead: A low barrier, typically about four feet high. There were several bulkheads in the long connecting drift between the Barnes-Hecker and Morris Mines, and one of the bulkheads served as a low dam to contain the worst of the water, mud, and debris that came from the Barnes-Hecker.

Cage: elevator in a mine shaft. Records show that for shift changes, it took 30 minutes to transport men up and down the shaft at Barnes-Hecker.

Caisson Method: There were mines in the North Lake and Gwinn Districts with shafts that were sunk through swamps. CCI contracted with a company called the Foundation Company to construct large vertical concrete cylinders through which earth could be removed while keeping groundwater at bay so the concrete shaft could be built. Caissons are also used during construction of the anchor supports for bridges. Interested readers can find additional information online.

CCI: Cleveland Cliffs Inc. is a vertically integrated steel and iron ore company that operates from the mining of raw materials to the manufacturing of finished steel products. They are a leading North American steel producer, particularly focused on value-added sheet products for the automotive industry.

District: CCI had groups of mines called districts overseen by superintendents. In addition to the North Lake District, there was a Gwinn District, Negaunee District, and Ishpeming District, to name a few.

Drift/drifting: Horizontal passageway. A drift can be large enough for equipment or small enough for a man to crawl through. The latter were nicknamed, "dog drifts." Drifting is the process of opening up a drift.

Dry building or dry: facility with lockers, showers, and individually assigned baskets and hooks for hanging wet mine clothing. Wet clothes were hoisted up near the ceiling to dry. Over time, CCI records show that first-aid rooms were added to the dry. The first-aid room was painted white, had an exam table, and was equipped with first-aid equipment. Ever since 1913, CCI conducted regular training in first aid at all mines, and classes in first aid were provided as part of a 1914-15 curriculum for men interested in advancement to supervisory positions.

Dry man: Person on duty in the dry. According to the Barnes-Hecker payroll records, dry man Jack Wills also served a few days as a pump helper.

Footage: Men with the title of "miner" were paid by the volume of ore, or footage, they produced. Footage goals were set by the shift bosses and/ore mine captains. The more ore that a crew produced, the more they were paid. Over the years, the safety department records state that the men's desire for higher pay sometimes took precedence over safety rules, resulting in higher rates of injury.

Headframe: a structural frame erected over a mine shaft, serving to support the hoisting machinery and cables used to lift miners and ore from the mine. The tower structure that facilitates the movement of materials and personnel between the surface and the underground.

Level: A main horizontal passageway with railroad tracks, haulage motors, and overhead electrical wires to which the motors were connected. Smaller passageways called drifts could branch off the main level. Vertical passageways are called raises. Small versions of levels or layers called sublevels were opened up to access the ore.

Location: Neighborhood of housing, often consisting of residences owned by mining companies, to provide living quarters for mine employees and their families. As an example, the Barnes-Hecker Mine was seven miles from the town of Ishpeming, so CCI built a Location of two rows of duplexes, a captain's house, and a one-room school. The Locations were within walking distance of the mines where the men worked. Rent for a Barnes-Hecker duplex was $5 a month, deducted from the man's paycheck.

Miner: Although "miner" is used in general terms to describe people employed in mining, a "miner" is a specific job title for a man who supervised small crews of about four men, assigned to a specific "contract," and paid by the amount of ore produced, rather than earning a fixed rate called a "company account." Miners were among the highest-paid employees.

Raise: a vertical passageway within a mine. Some raises had specific uses. A "man road" contained ladderways for climbing up and down. A skip road was dedicated to the large buckets used to lift ore.

Scraper: Horizontal bucket attached to heavy chains that pulled blasted ore to chutes where ore fell into trammers' carts or into ore cars attached to motors on railroad tracks.

Shaft: The main vertical passageway into a mine.

Skip/skip tender: A skip was like a bucket that carried ore up the shaft. A skip tender was stationed at the shaft to make sure skips ran smoothly. Skip tenders also oversaw "skip pits" where excess rock or ore could overflow by the shaft.

Stemmer: A stemmer was a man who could be placed wherever he was needed. Walter Tippett's employment slip identifies him as a stemmer; his accident report lists him as a miner. Tippett had previously worked at the Morris-Lloyd/North Lake Mine, and in the White Pine Copper Mine.

Sub-level or sub: like a partial floor of a house, a sublevel is a layer or horizon between main levels. Sub-levels could be irregular because they followed the ore deposits.

Trammer: This is the job title of men who pushed heavy carts filled with ore to the main levels where the carts could be emptied into rail cars for transport to the shaft.

Vug: A vug is an underground watercourse, where large amounts of water could be found. Vugs can exist in cracks and crevices that can run for miles. The 1926 CCI annual report stated that the most likely cause of the Barnes-Hecker Disaster was that a drill had "holed into a vug."

Bibliography

"Coroner's Inquest into the deaths of William E. Hill, Arvid Wepsala, Henry Haapala, Nels Hill, Thomas Kirby Jr., Thomas Kirby Sr., William F. Tippett, Joseph Mankee, William Huot and John J. Hannah (sic), State of Michigan, County of Marquette, Case no. 1528, Ishpeming MI, Feb. 2, 1927.

"Glossary of Mining Terms," https://www.sec.gov/Archives/edgar/data/1165780/000116578003 000001/glossary.htm, retrieved on several dates 2021-2024.

Adgate, Frederick W. "Sinking reinforced concrete shafts through quicksand," *Proceedings of the Lake Superior Mining Institute, XIV,*" pp. 55-70, *Iron Ore,* Ishpeming, MI Sept 4, 1909.

Anderson, Douglas, correspondence re: William and Clara Huot with Mary V. Tippett, 2017-2019.

Anderson, William and Anderson, Peter, interview re: Peter and Evangeline DeRoche, with Mary V. Tippett, Negaunee, MI, Nov. 9, 2018.

Armstrong, Amy comp., *1913 Harvard University Alumni Directory* [database online], Provo, UT, USA, Ancestry.com Operations, Inc., 2001. Original data: Harvard University Alumni Association, *Harvard University Directory 1913,* Boston Ma, USA, Harvard University Press, 1913, retrieved Oct. 15, 2020.

Arsenault, Sandra (former owner of the Gossard Building), correspondence with Mary V. Tippett, Feb. 7, 2022.

Bartanen, Carole Maki and Bartanen, Bruce. Interview re: Emil and Lempi Maki with Mary V. Tippett, Negaunee, MI, Oct. 16, 2019.

Bitner, Susan, and Marietti, Cheryl, photograph at grave of Henry Huot in Belgium.

Borden, Mary Trudell. interview with Julie Mosurak re: Edward Trudell, Sept. 28, 2020.

Bowers, Lisa. "From tragedy to tradition: Lampshire-Ellersick descendant worked below the Barnes-Hecker Mine," *The Mining Journal,* Oct. 31, 2016.

Bowers, Lisa, "Through a child's eyes," *The Mining Journal,* Marquette, MI, Oct. 24, 2016.

Boyum, Burton H. *The saga of underground mining,* John M. Longyear Research Library, (Marquette MI), 1983.

Britton, Doreen. correspondence re: Kirbys with Mary V. Tippett, 2017-2020.

Burt, John and Burt, William A., Survey Map, Twp 47 N, R 28 W, Archives of Michigan, 1846 & 1848.

Bushong, Diana Warner and Bushong Lloyd; Warner, LeRoy and Warner, Robert. Interview re: Nick and Mary Valenti, with Mary V. Tippett, Negaunee, MI, Sept. 20, 2019.

Chicago Daily Tribune, "160 orphaned in disaster in Michigan mine; death list 51...," Nov, 5, 1926.

Chicago Herald and Examiner, "Relief work futile," Newspaper photograph," Nov. 1926, private collection.

Chicago Evening American, undated photo, courtesy of Don Hill.

Chipman, Roger. Interview re: Solomon and Minnie MIllimaki, with Mary V. Tippett, Negaunee, MI, Oct. 14, 2019.

Chrisman, Sonya and Turner, Bruce. "Barnes-Hecker: Memories of a Misfortune," WNMU Public TV13, Marquette, MI, 2001.

City of Ishpeming, "Cemetery Burial Records," MI, www.ishpemingcity.org, numerous retrieval dates.

Cleveland Cliffs Iron Company, *Report of accident or injury to employe (sic),* Nos. 142-192, 195, Barnes-Hecker Mine, Leo LaFond Collection, Cliffs Shaft Mine Museum, Ishpeming MI, Nov. 4, 1926.

Cleveland Cliffs Iron Company, *Barnes-Hecker Mine payroll, 1921-1927,* (two volumes) Central Upper Michigan and Northern Michigan University Archives, Marquette MI.

---Education department annual reports, 1914-15, Central Upper Peninsula and Northern Michigan University Archives, Marquette, MI, retrieved April 30, 2017.

---Negaunee Mine annual report, pp. 83-86.

---Annual reports, departmental reports and maps, 1909-1927, Central Upper Peninsula and Northern Michigan University Archives, Marquette MI. (Multiple retrieval dates 2016-2024)

---Pension Department annual report 1927, Central Upper Peninsula and Northern Michigan University Archives, Marquette, MI, retrieved April 28, 2017, p. 500.

---Pension Department Annual Report, 1928, p. 421, Central Upper Peninsula and Northern Michigan University Archives, retrieved April 28, 2017.

---*Pension Department Annual Report,* Dec. 1926, pp. 103-107, collection of Tippett descendants.

---Pension department annual reports 1926-1932, retrieved on multiple dates, 2016-2024.

Conibear, William Sr., Find A Grave Memorial ID 185103382 for William Conibear Sr., retrieved July 23, 2022.

Conibear, William, "Development of safety in the iron mines of Michigan," *Proceedings of the Lake Superior Mining Institute,* 1936, p. 24.

Coon, Georgann Kippola, "Marriage License of Rita Kirby Cowling and Charles Cowling," Aug. 14, 1922.

--- interview re: Kirby and Cowling families with Mary V. Tippett, May 4, 2021, 2022-2024.

Crump, R.M., "Memorial to Leslie Park Barrett, 1887-1972," leaflet, The Geological Society of America, Boulder Co., 1972, courtesy Marquette Regional History Center.

Decuyper, Ronny, "Adopt a U.S. Tommy Program," correspondence re: Huot family with Mary V. Tippett, 2017-2018.

DeRoche Peter and Evangeline, "Marquette County Marriage Record, fourth quarter of 1911," Ancestry.com, retrieved June 24, 2021.

DiBernardo, Angelo, unpublished essay on the lives of shift boss Sam Phillippi and Mary Phillippi (Filippi), Private collection, 1946.

Draffen Robert, and Draffen, Michael, telephone interview re: Chapman family with Mary V. Tippett, July 22, 2022.

Finnish American Heritage Center, [formerly Finlandia University] Archives. Correspondence re: Finnish accordion/zither, with Mary V. Tippett, Hancock, MI, Feb 11, 2022.

Friggens, Thomas, *No Tears in Heaven: the 1926 Barnes-Hecker Mine Disaster,* Michigan Historical Center, Michigan Department of State, Lansing, MI, 1998.

Gravedoni, Margaret. Correspondence re: Ted Kiuru with Mary V. Tippett, 2016-2020.

Heino, Arvid and Tyyne Alto (Aalto), *Marriage license,* Marquette County, MI, Nov. 1, 1926, Ancestry.com, retrieved Sept. 30, 2020.

Heino, Paul. interview re: Heino family with Mary V. Tippett, Sept. 29, 2020.

Hooper, Mary Warner and Hooper, Robert. Telephone conversations and correspondence re: Nick and Mary Valenti and monument on capped shaft with Mary V. Tippett, 2019-2020.

Hosang, Carl. Barnes-Hecker Remembrance Committee 2016-2019.

Howe, Manthel, "Men must work and women weep – in Mining," *The Daily Mining Journal,* Marquette, MI, Nov. 5, 1926.

Hunt, Al, "Mine disaster changed Clifford's life," *The Mining Journal,* Marquette, MI, Nov. 6, 1988, courtesy Clifford Trudell.

Hunt, Al, "Monument to 1926 cave-in moved to museum," *The Mining Journal,* Marquette MI, Sept. 18, 1988.

Huot, William. Michigan Department of Community Health, Division for Vital Records and Health Statistics; Lansing, MI, *Death records,* Ancestry.com, retrieved Sept. 19, 2020.

Iron Ore, "At North Lake," Ishpeming MI Aug 28, 1909.

Iron Ore, "Barnes-Hecker making headway," Ishpeming, MI, Aug. 31, 1918.

Iron Ore, "Cave-in at Barnes-Hecker costs fifty-one lives," Ishpeming, MI, Nov. 6, 1926.

Iron Ore, death notice of Jackie Hanna [untitled], Ishpeming, MI, Aug. 26, 1939.

Iron Ore, "Idled Barnes-Hecker headframe will go," Ishpeming, MI, Nov. 8, 1952.

Iron Ore, "Minutes of the Marquette County Board," Ishpeming, MI, Nov. 27, 1926.

Iron Ore, "Recover three more bodies: Remains of Jack Hanna, Thos. Kirby Jr., and Jos. Mankee out of Mine," Ishpeming, MI, Nov, 13, 1926.

Iron Ore, "The Mine Disaster," *Ishpeming*, MI, Nov. 6, 1926.

Isherwood, Shelly Maki. Correspondence re: Emil and Lempi Maki with Mary V. Tippett, March 21, 2018.

Jackson, Chas. F. "Mining by the Top-Slicing Method," www.OneMine.com retrieved Sept. 9, 2022.

Kallio, David, interview re: Arvid and Sanna Kallio with Mary V. Tippett, Negaunee, MI, April 4, 2019.

Kasper, Karen Johnson, Correspondence with Mary V. Tippett, Feb. 5th, 2022.

Koskinen, Karen, Koskinen Brenda, and Reed, Erika. Interview re: Uno and Vieno Koskinen with Mary V. Tippett, Negaunee, MI, Aug. 9, 2019.

Lake Superior Iron Ore Association, *Lake Superior iron ores; Mining directory and statistical record of the Lake Superior iron ore District of the United States and Canada, 2nd ed.*; Cleveland OH, 1952, p. 76, collection of the Michigan Iron Industry Museum.

Lee, William H. A history of Phosphate Mining in Southeastern Idaho, U.S. Geological Survey, p.69, Open file report 00-425 retrieved from https://pubs.usgs.gov/of/2000/of00-425/chapters/01-02_Title_TOC.pdf on May 31, 2025

Lucas, Thomas. Interview re: Richard and Helen Lucas, with Mary V Tippett Negaunee, MI, Oct. 25.2018.

Makinen, Patricia Hill. Interview re: William and Elizabeth Hill with Mary V. Tippett, Negaunee MI, Aug. 13, 2019.

Marietti, Cheryl and Bitner, Susan. Correspondence re: Huot family with Mary V. Tippett, April, 2024.

Marquette County Clerk, "Marriage License of Mae Hill and Wilfred LaForge," Ancestry.com, retrieved June 21, 2021.

Mattson, Anna Hill, correspondence re: Nels and Mae Hill with Mary V. Tippett, June 2022.

Michigan Department of Health, Arvid Heino, "Certificate of death," December 31, 1926, Ancestry.com., retrieved Sept. 13, 2020.

Michigan Iron Industry Museum, "Barnes-Hecker Mine Memorial Rededication," program, Negaunee, MI, June 11, 1989.

Michigan Iron Industry Museum, Untitled video, rededication of Barnes-Hecker Monument, June 11, 1989.

Millimaki, Gerald. Telephone interview with Mary V. Tippett, March 14, 2021.

Mosurak, Julie, correspondence re: Edward Trudell with Mary V. Tippett, Sept. 2020.

Negaunee Iron Herald, William Conibear [Sr.] untitled death notice, Negaunee, MI, July 7, 1887.

Nelson, Frances (Stakel), Nelson, Bruce K., Williams, Robert W. eds. *Memoirs of Charles J. Stakel, Cleveland Cliffs Mining Engineer, Mine Superintendent, and Mining Manager,* The John M. Longyear Research Library, Marquette County Historical Society, Marquette MI, 1994, p. 99.

Newett, George O., "The Marquette Iron Range," *Proceedings of the Lake Superior Mining Institute,* Vol. XXIV, Aug. 13-14, 1925, pp. 10-21, retrieved Sept. 27, 2022.

Nyyssölä, Suvi, email correspondence re: Erkki and Amalia Timoharju with Mary V. Tippett, 2019-2020.

Perala, Gary and Perala, Ronald (Jack), correspondence re: Lampshire brothers with Mary V. Tippett, 2016-2020.

Peterson, John A. and Rose; City of Ishpeming, Michigan burial records, retrieved Sept. 8, 2020.

Peterson, John Paul, obituary, bjorkandzhulkie.com, retrieved Sept 8, 2020.

Proceedings of the Lake Superior Mining Institute, William Kelly Obituary. Vol. XXX, Obituary, Sept. 14-15, 1939, p. 38.

Rönkkö, Raili. Email correspondence re: Erkki and Amalia Timoharju with Mary V. Tippett, 2019-2020.

Schindler, Elaine Ranta. Correspondence re: Elias and Amanda Ranta with Mary V. Tippett, 2017-2020.

Smail, Dawne Tippett, Interview with Shelly Maki Isherwood, *Upper Michigan Today*, WLUC-TV6 re: Rededication of Barnes-Hecker Monument at Michigan Iron Industry Museum, Negaunee, MI, June 1988, videotape, courtesy of Melody Smail Visser.

Snell, Constance Millimaki. Interview re: Solomon and Wilhelmina MIllimaki with Mary V. Tippett, April 7, 2021.

Soria, Merja. "World Music Instrument: Kantele," www.centerforworldmusic.org, retrieved Feb. 11, 2022.

Soyring, John A. Email correspondence re: Erkki and Amalia Timoharju with Mary V. Tippett, 2019-2020.

Sromalski, Kate Trebilcock. Correspondence re: Olaf Trondson with Mary V. Tippett, 2016-2020.

Stoping Mining Methods – Metallurgist and Mineral Processing,"
    metallurgist.com, retrieved Sept. 9, 2022.

The Daily Mining Journal, "Five funerals today," Marquette, MI, Nov. 5,
    1926.

The Daily Mining Journal, "CCI not to blame," Marquette, MI, Nov. 8,
    1926.

The Daily Mining Journal, "CCI pioneer in mine safety," Marquette, MI,
    Nov. 8, 1926.

The Daily Mining Journal, "CCI to abandon Barnes-Hecker; Danger to
    men too great, say officials," Marquette MI, Jan. 12, 1927.

The Daily Mining Journal, "Chief of Police Martin Ford and patrolman
    Thornton killed in gun fight," Marquette MI, Aug. 25, 1924.

The Daily Mining Journal, "Coroner's inquest returns open verdict,"
    Marquette, MI, Feb. 3, 1927.

The Daily Mining Journal, "Find another body in Mine: Corpse of Joe
    Mankee recovered from shaft, work speeded up," Marquette, MI,
    Nov. 13, 1926.

The Daily Mining Journal, "Find more bodies in Mine, Marquette, MI,
    Nov. 8, 1926

The Daily Mining Journal, "Find two more bodies in Barnes-Hecker,
    Have reached all in shaft, belief," Marquette MI, Nov. 10, 1926.

The Daily Mining Journal, "Funeral on Thursday for W.W. Graff;"
    Marquette, MI, Oct. 4, 1944, courtesy Marquette Regional
    History Center.

The Daily Mining Journal, "Heart attack fatal to S.R. Elliott, Executive of
    CCI for many years; Dies at home in Ishpeming this morning,"
    Marquette, MI, April 13, 1946, courtesy Marquette Regional
    History Center.

The Daily Mining Journal, "Ishpeming briefs, Birth of baby boy to
    Florence and Jack Hanna," Marquette, MI, Nov 17, 1926.

The Daily Mining Journal, "Large number of cars in funeral cortege when
    disaster victim buried; Many paid tribute to Nels Hill,"
    Marquette, MI, Nov. 10, 1926.

The Daily Mining Journal, "Mat Clash at Liberty Hall Friday night,"
    [Walter Tippett] Marquette MI, Sept. 13, 1926, p. 13.

The Daily Mining Journal, "Rush work at flooded mine," Marquette, MI,
    Nov, 5, 1926.

The Daily Mining Journal, "Speed work at flooded mine: make progress," Nov. 6, 1926, p. 13.

The Daily Mining Journal, "Veteran Mine Captain Dies at His Home," [Allyvion Miners], Marquette, MI., May 2, 1940, Courtesy Marquette Regional History Center.

The Daily Mining Journal, "Water rises eight feet an hour in shaft," Marquette, MI, Nov. 20, 1926.

The Daily Mining Journal, Ishpeming and environs, Chapman son leaving for Pennsylvania, Marquette, MI, Nov, 20, 1926.

The Daily Mining Journal, "City snow plows begin their work," Marquette MI, Nov 10, 1926.

The Mining Journal, "Barnes-Hecker Disaster relived, Marquette, MI, Aug 26, 1972.

The Mining Journal, "Bergdahl dies at 97; Rites on Tuesday," Marquette, MI, May 6, 1946, courtesy Marquette Regional History Center.

The Mining Journal, "Capt. Nault dies after long illness," Marquette, MI. Dec. 11, 1943, Courtesy Marquette Regional History Center.

The Mining Journal, "David Crago dies after long illness," Marquette, MI, Oct. 30, 1933, p. 10, courtesy Marquette Regional History Center.

The Mining Journal, "Death takes J.W. Skewes in Ishpeming," *Marquette, MI, Jan. 18, 1961,* courtesy Marquette Regional History Center.

The Mining Journal, "William Connibear [sic] first mine safety director, dies," Marquette, MI, Feb. 12, 1956, Courtesy Marquette Regional History Center.

The Mining Journal, James Murphy obituary, *Marquette, MI, April 13, 1931,* courtesy Marquette Regional History Center.

The Mining Journal, John Webster obituary, Marquette, MI, Jan. 17, 1952, courtesy Marquette Regional History Center.

The Mining Journal, William Rashleigh obituary, Marquette MI, July 26, 1957, courtesy Marquette Regional History Center.

The Mining Journal, Olaf Trondson obituary, Marquette, MI, Aug. 22, 1979, courtesy Marquette Regional History Center.

The Mining Journal, "Tragic heirloom," [Kirby pocket watch gifted] Marquette, MI, Nov. 3, 2008, p. 6A.

Tippett, Annie, "Stories of Grandma Hill," re: Walter Tippett, unpublished, Feb. 8., 2009.

Tippett, Eric William, Telephone interview re: Wm. and Nellie Tippett with Mary V. Tippett, March 24, 2019.

Tippett, T. Wilfred, "Recollections, Barnes Hecker Disaster, Barnes Hecker Shaft House on Surface," unpublished manuscript, Leo LaFond Collection at Cliffs Shaft Mine Museum, May 23, 1989.

Tippett, Walter R. "How the Barnes-Hecker Tragedy affected the Tippett Family," unpublished manuscript, Private collection, 2008 rev. 2014.

---"Notes on the life of Walter Tippett," unpublished manuscript, private collection, 2008 rev. 2019.

Trondson, Olaf, U.S. Department of Veterans Affairs BIRL death file 1850-2010; retrieved from Ancestry.com Sept. 11, 2020.

Trudell, Clifford Jr., telephone interviews with Mary V. Tippett, 2016-2020.

U.S. Department of Agriculture Weather Bureau, weather ledger for November, 1926.

U.S. Department of Commerce – Bureau of the Census, *Fourteenth Census of the United States: 1920 – Population*; Ely Township, Marquette County Michigan, sheets 8A, 8B, 9A, courtesy of Cliffs Shaft Mine Museum, Ishpeming, MI.

--- *Fifteenth Census of the United States: 1930*, Population Schedule, Ishpeming, MI, Ward 8, Sheet 2 (Ancestry source citation as Year: 1930; Census Place: *Ishpeming, Marquette, Michigan; Page 2A;* Enumeration District: 0014; FHL Microfilm 2340745).

Wallberg, Gardner A. "The Barnes-Hecker Disaster," unpublished manuscript, Ishpeming, MI, Sept. 1972. P.8.

Warlin, Jean. Obituary, Bjork and Zhulkie Funeral Home, Ishpeming, MI, Dec. 17, 2019, retrieved Aug. 15, 2020.

Williams, D.J., "Lambeau's Packers crush Ishpeming in 'rough contest,'" [Green Bay Packers vs. Ishpeming-Negaunee Allstars in 1919], *Porcupine Press,* Trenary, MI, July 28, 1993. [The article cites *The Green Bay Press Gazette and The Marquette Mining Journal,* Oct. 20, 1919, and the Ishpeming *Iron Ore,* Oct. 25, 1919.]

WLUC-TV 6, Barnes-Hecker Monument arrives at Michigan Iron Industry Museum, Marquette, MI, Sept, 15, 1988, videotape, collection of Melody Smail Visser.

WLUC-TV 6, Rededication of Barnes-Hecker Monument at Michigan Iron Industry Museum, June 11, 1989, videotape, collection of Melody Smail Visser.

About the Author

Mary Tippett is a granddaughter of Walter Tippett, who perished in the Barnes-Hecker tragedy. In 2016, she led the 90th Anniversary Barnes-Hecker Remembrance Project, and continues to manage the Facebook page with a following of more than 850 people. She currently serves on the board of the Friends of the Michigan Iron Industry Museum in Negaunee.

In 1994, she conducted oral histories with underground miners of her Dad's generation, and all materials are in the Central Upper Peninsula and Northern Michigan University Archives available online.

The Tippett family genealogy has been traced back to the mid-1600s in Cornwall, England, with the first documentation of mining in the 1860s. Both the Finnish and Cornish sides of Mary's extended family including her Dad, brother, and son-in-law are among several generations of mining people.

Mary is retired from the position of Senior Director of Marketing and Community Relations at Marquette General Health System (now U.P Health System), where she was also chief information officer during two events that drew national media attention. Mary was previously a writer/producer of local television commercials at WLUC-TV6.

She holds a BA, Summa cum Laude, from Northern Michigan University, and was inducted into Phi Kappa Phi National Honor Society. She has been listed in *Who's Who of American Women*, *Who's Who in the Media and Communications*, and *Who's Who in the Midwest*. She has also received recognition for writing/graphic design work and special programs/events in the community.

She served on the board of the Economic Club of Marquette County; was the first female chair of the congregation council of Messiah Lutheran Church; vice-chair of Friends of the DeVos Art

Museum at Northern Michigan University; board member of the U.P. Children's Museum; chair of the Title One Parent Advisory Committee in the Royal Oak, Michigan, school system; and served on numerous other advisory committees and boards since the 1970s. In 1982, she initiated the effort to start the After School Program in the Marquette Area Public Schools and served on its board. The program has since grown and changed, but still serves school-aged children throughout Marquette County.

She is a native of Michigan's Upper Peninsula, and has lived most of her life within 20 miles of the Barnes-Hecker Mine site.

Appendix - Map

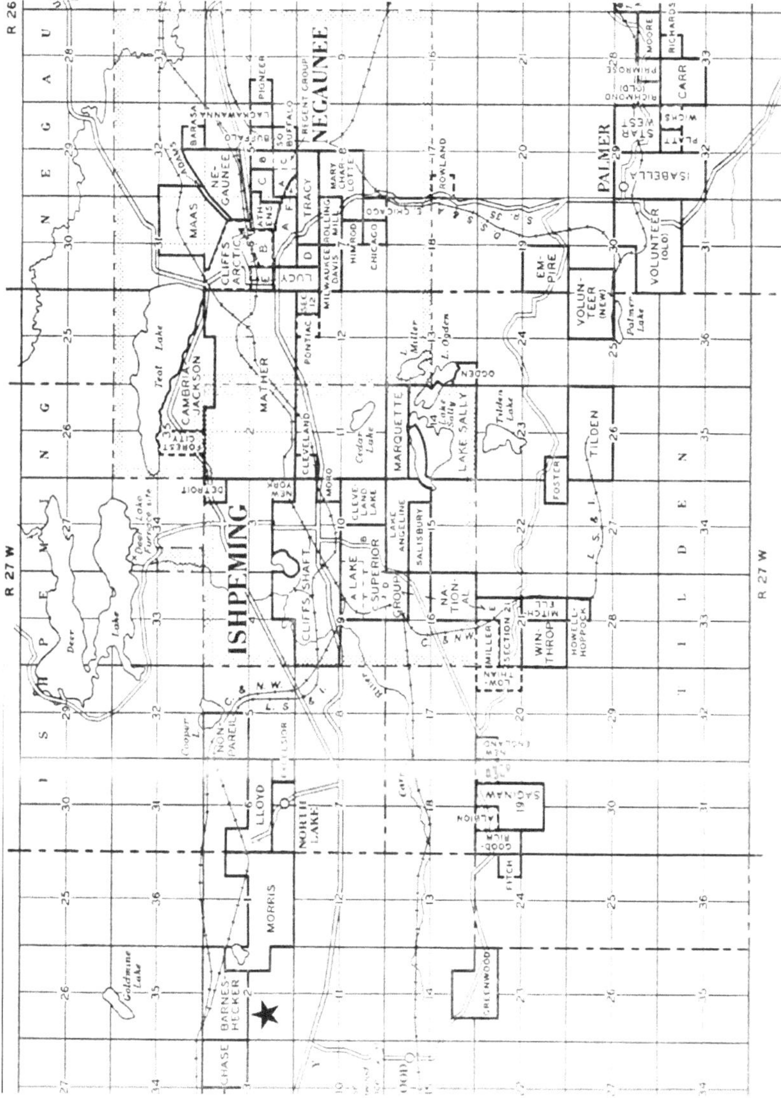

This map shows locations of mines in Ishpeming, Negaunee, and Palmer. The book has been digitized and is in the public domain. It includes all productive iron ore mines in the Lake Superior District until 1952.
From *Lake Superior Ores, Mining Directory and Statistical Record of the Lake Superior Iron Ore District of the United States and Canada, 2nd Ed.*, Lake Superior Iron Ores Association, Cleveland, Ohio (1952).
Courtesy of the Michigan Iron Industry Museum.